AF690998

BRILJENT

RFID+

Elaine M. Cooney

Professor of Electrical and Computer Engineering Technology

CAPE CORAL HIGH SCHOOL

The Complete Review of Radio Frequency Identification

THOMSON

DELMAR LEARNING™

Australia Canada Mexico Singapore Spain United Kingdom United States

THOMSON

DELMAR LEARNING

RFID+: The Complete Review of Radio Frequency Identification

Elaine M. Cooney
Research: Mike Gifford and Adam Wickman

Briljent LLC
President: Kathy Carrier
Editing: Brian Foss, Ann Smith, Jill Starbuck
Production and Design: Steve Pepple

Vice President, Technology and Trades SBU:
David Garza

Editorial Director:
Sandy Clark

Senior Acquisitions Editor:
Stephen Helba

Marketing Director:
Diane Yarnell

Senior Channel Manager:
Dennis Williams

Marketing Coordinator:
Stacey Wiktorek

Senior Production Manager:
Larry Main

Art/Design Coordinator:
Francis Hogan

Senior Editorial Assistant:
Dawn Daugherty

COPYRIGHT © 2006 Indiana University – Purdue University Indianapolis. Thomson, the Star Logo, and Delmar Learning are trademarks used herein under license.

Printed in the United States of America
1 2 3 4 5 XX 07 06 05

For more information contact
Thomson Delmar Learning
Executive Woods
5 Maxwell Drive, PO Box 8007,
Clifton Park, NY 12065-8007
Or find us on the World Wide Web at
www.delmarlearning.com

ALL RIGHTS RESERVED. No part of this work covered by the copyright hereon may be reproduced in any form or by any means—graphic, electronic, or mechanical, including photocopying, recording, taping, Web distribution, or information storage and retrieval systems—without the written permission of the publisher.

For permission to use material from the text or product, contact us by
Tel. (800) 730-2214
Fax (800) 730-2215
www.thomsonrights.com

Library of Congress Cataloging-in-Publication Data:
Card Number:

ISBN: 1-4180-5231-1

NOTICE TO THE READER

Publisher does not warrant or guarantee any of the products described herein or perform any independent analysis in connection with any of the product information contained herein. Publisher does not assume, and expressly disclaims, any obligation to obtain and include information other than that provided to it by the manufacturer.

The reader is expressly warned to consider and adopt all safety precautions that might be indicated by the activities herein and to avoid all potential hazards. By following the instructions contained herein, the reader willingly assumes all risks in connection with such instructions.

The publisher makes no representation or warranties of any kind, including but not limited to, the warranties of fitness for particular purpose or merchantability, nor are any such representations implied with respect to the material set forth herein, and the publisher takes no responsibility with respect to such material. The publisher shall not be liable for any special, consequential, or exemplary damages resulting, in whole or part, from the readers' use of, or reliance upon, this material.

Contents in Brief

Table of Contents

INTRODUCTION

Introduction

Who Is CompTIA?

CompTIA is a nonprofit global trade association headquartered in Chicago, Illinois. For 23 years, CompTIA has been dedicated to advancing the growth of the information technology (IT) industry, and has more than 20,000 members in 102 countries representing the international technology community. CompTIA's goal is to provide a unified voice, global advocacy, and leadership in the effort to advance IT industry growth by promoting industry standards, professional competence, education, and business solutions.

What Is RFID+ Certification?

The CompTIA RFID+ certification is an industry accepted, vendor-neutral certification program designed to provide a standard by which to validate an RFID technologist's knowledge and skills. Topics covered by the certification include RFID installation, maintenance, and hardware and software repair. In February 2005, CompTIA conducted a survey of 274 IT managers, and found that more than 77 percent believed that certification is an essential step in validating employees' skills. This survey also found that 79 percent of managers believe that lack of certification is a gap in an employee's skill set. The RFID+

certification was developed to address the growing need for skilled RFID professionals by creating a standard means to gauge RFID expertise.

Why Get Certified?

Certification is a valuable component of success in any field, but it is especially important in technological fields. Certification not only makes professionals more marketable for future opportunities, it ensures that they stay current with the latest technology and achieve the expertise necessary to succeed within their field. The CompTIA RFID+ certification is a practical way for technologists to measure their knowledge and experience in RFID technology. For companies and other institutions, RFID+ certification is a promising way to ensure employee competence.

Why Use This Book?

As a result of mandates by the likes of Wal-Mart and the U.S. Department of Defense that their vendors (which number in the thousands) adopt RFID technology, RFID has moved to the forefront of supply chain management and logistics. The RFID market is expanding rapidly, and can only be expected to continue its growth in the future, which presents strong prospects for technical professionals to build careers in the RFID field. Certification is one of the surest ways for an individual to validate his or her expertise and advance in this field. This book is designed to help prepare the reader for the CompTIA RFID+ certification exam and to provide valuable insight into essential RFID concepts.

How to Use This Book

The primary objective of *RFID+: The Complete Review of Radio Frequency Identification* is to provide the most up-to-date information about existing RFID systems as it relates to success on the certification exam. The reader will be guided step-by-step through a series of chapters related to RFID technology and the CompTIA certification exam. While some highly technical information regarding electronic and radio frequency technology is presented throughout the course of this book, this information is presented in a straightforward and logical manner with sections that explain the more complex topics. The preliminary chapters of this book explain the basic engineering concepts in RFID and the main components of an RFID system, covering everything from operating standards to actual system operation. The final chapters deal with regulations, security issues, and some practical applications that are currently used in the field. All of the information provided in this book is arranged to offer the reader the best possible preparation for the actual CompTIA certification exam.

Chapter Objectives

Each chapter of the book contains features that aid studying for the CompTIA Certification Exam.

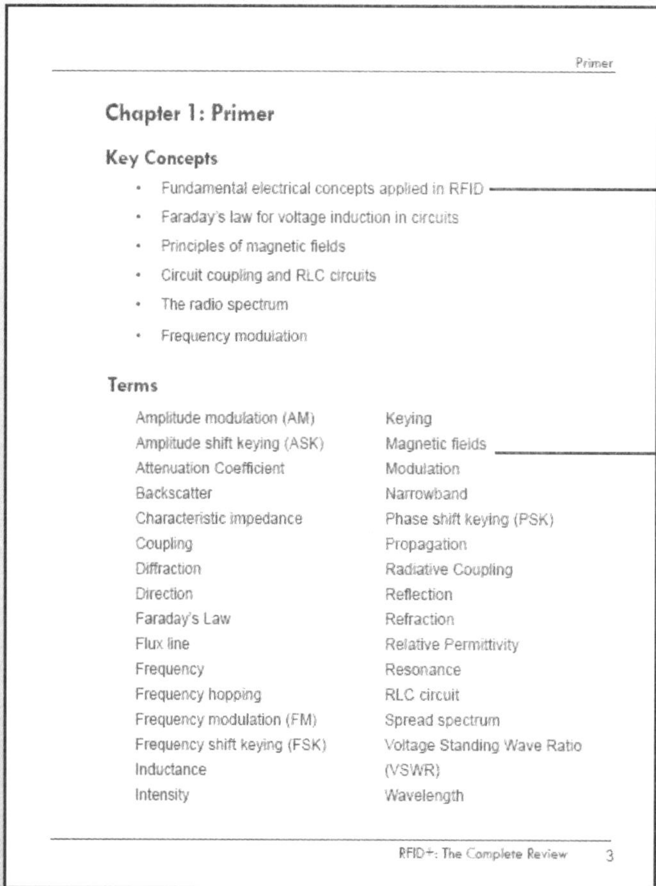

Primer

Chapter 1: Primer

Key Concepts

- Fundamental electrical concepts applied in RFID
- Faraday's law for voltage induction in circuits
- Principles of magnetic fields
- Circuit coupling and RLC circuits
- The radio spectrum
- Frequency modulation

Terms

Amplitude modulation (AM)	Keying
Amplitude shift keying (ASK)	Magnetic fields
Attenuation Coefficient	Modulation
Backscatter	Narrowband
Characteristic impedance	Phase shift keying (PSK)
Coupling	Propagation
Diffraction	Radiative Coupling
Direction	Reflection
Faraday's Law	Refraction
Flux line	Relative Permittivity
Frequency	Resonance
Frequency hopping	RLC circuit
Frequency modulation (FM)	Spread spectrum
Frequency shift keying (FSK)	Voltage Standing Wave Ratio (VSWR)
Inductance	Wavelength
Intensity	

RFID+: The Complete Review 3

Key Concepts
Each chapter begins with a list of the concepts that will be discussed in the chapter.

Terms
Each chapter contains the key terminology for the chapter.

Instructional Features

At the end of each chapter is a review and discusion of the chapter's material. The multiple choice questions are representative of the type of questions that will appear on the RFID+ exam. The discussion questions provided will reinforce the reader's understanding of the concepts and scenarios that take precedence on the exam. Answers to review questions are provided at the end of the book.

Primer

Review

Questions

Answers are provided in

1. What units are used to measure inductance?
   a. Ohms
   b. Volt
   c. Watt
   d. Henries
2. Which will have a longer wavelength?
   a. signal of 10 MHz
   b. signal of 50 MHz
   c. signal of 100 MHz
   d. signal of 200 MHz
3. The correct relationship between period and frequency is?
   a. f = 1/T
   b. T = 1/F
   c. fT = 1
   d. all of the above
4. In an inductor, what is the cause of opposition to the circuit?
   a. reluctance
   b. resistance
   c. capacitance
   d. none of the above

Practice Exam Questions

RFID+: The Complete Review 65

Summary
Each chapter ends with a summary of the chapter's discussion.

Review Questions
Multiple choice and discussion questions are provided at the end of each chapter. The questions are directly based on the CompTIA RFID+ objectives, and test the reader's comprehension.

The Exam Objectives

The following is an outline of the eight domains and domain objectives that CompTIA has identified for the RFID+ certification exam. This outline is also available at the CompTIA Web site, www.comptia.org

Domain 1.0 Interrogation Zone Basics

1.1 Describe interrogator functionality

- 1.1.1 I/O capability
- 1.1.2 Handheld interrogators
- 1.1.3 Vehicle mount interrogator
- 1.1.4 LAN/Serial communications
- 1.1.5 Firmware upgrades
- 1.1.6 Software operation (GUIs)

1.2 Describe configuration of interrogation zones

- 1.2.1 Explain interrogator to interrogator interference
- 1.2.2 Optimization
- 1.2.3 System performance and tuning
- 1.2.4 Travel speed and direction
- 1.2.5 Bi-static/monostatic antennas

1.3 Define anti-collision protocols (e.g., number of tags in the field/response time)

1.4 Given a scenario, solve dense interrogator environment issues (domestic/ international)

1.4.1 Understand how a dense interrogator installation is going to affect network traffic

1.4.2 Installation of multiple interrogators, (e.g., dock doors, synchronization of multiple interrogators, antenna footprints)

Domain 2.0 Testing and Troubleshooting

2.1 Given a scenario, troubleshoot RF interrogation zones (e.g., root-cause analysis)

2.1.1 Analyze less than required read rate

2.1.1.1 Identify improperly tagged items

2.1.2 Diagnose hardware

2.1.2.1 Recognize need for firmware upgrades

2.1.3 Equipment replacement procedures (e.g., antenna, cable, interrogator)

2.2 Identify reasons for tag failure

2.2.1 Failed tag management

2.2.2 ESD issues

2.3 Given a scenario, contrast actual tag data to expected tag data

Domain 3.0 Standards and Regulations

3.1 Given a scenario, map user requirements to standards

3.1.1 Regulations, standards that impact the design of a particular RFID solution

3.2 Identify the differences between air interface protocols and tag data formats

3.3 Recognize regulatory requirements globally and by region (keep at high level, not specific requirements—may use scenarios)

3.4 Recognize safety regulations/issues regarding human exposure

Domain 4.0 Tag Knowledge

4.1 Classify tag types

4.1.1 Select the RFID tag best suited for a specific use case

4.1.1.1 Pros and cons of tag types

4.1.1.2 Tag performance

4.1.1.2.1 Tag antenna to region/ frequency

4.1.2 Identify inductively coupled tags vs. backscatter

4.1.3 Identify the differences between active and passive

4.2 Given a scenario, select the optimal locations for an RFID tag to be placed on an item

Domain 5.0 Design Selection

5.1 Given a scenario, predict the performance of a given frequency and power (active/passive) as it relates to read distance, write distance, tag response time, storage capacity

5.2 Summarize how hardware selection affects performance (may use scenarios)

5.2.1 Antenna type

5.2.2 Equipment mounting and protection

5.2.3 Cable length/loss

5.2.4 Interference considerations

5.2.5 Tag type (e.g., active, passive, frequency)

Domain 6.0 Installation

6.1 Given a scenario, describe hardware installation using industry standard practices

6.1.1 Identify grounding considerations (e.g., lightning, ground loops, ESD)

6.1.2 Test installed equipment and connections (pre-install and post-install)

6.2 Given a scenario, interpret a site diagram created by a RFID architect describing interrogation zone locations, cable drops, device mounting locations

Domain 7.0 Site Analysis

7.1 Given a scenario, demonstrate how to read blueprints (e.g., whole infrastructure)

7.2 Determine sources of interference

7.2.1 Use analysis equipment such as a spectrum analyzer, determine if there is any ambient noise in the frequency range that may conflict with the RFID system to be installed

7.3 Given a scenario, analyze environmental conditions end-to-end

Domain 8.0 RF Physics

8.1 Identify RF propagation/communication techniques

8.2 Describe antenna field performance/characteristics as it relates to reflective and absorptive materials (may use scenarios)

2 0621 20071330 2

8.3 Given a scenario, calculate radiated power output from antenna based on antenna gains, cable type, cable length, interrogator transmit power (include formulas in scenario)

Domain 9.0 RFID Peripherals

9.1 Describe installation and configuration of RFID printer (may use scenarios)

9.2 Describe ancillary devices/concepts

9.2.1 RFID printer encoder

9.2.2 Automated label applicator

9.2.3 Feedback systems (e.g., lights, horns)

9.2.4 RTLS

CAPE CORAL HIGH SCHOOL

Objectives by Chapter

To best illustrate the relationship between concepts and components, the topics of this book are presented in a different order than the exam objectives. The following outline maps the topics of this book with CompTIA's RFID+ certification objectives.

Chapter 1: Primer	
Domains 4.0, 5.0, 8.0, 9.0	
Faraday's Law	Objective 4.1
Magnetic Fields	
Coupling	Objective 4.1
LC & AC Circuits	
Wavelength	
How Different Materials Affect RF	Objective 4.2 , 5.2
RF Transmission and Reception	Objective 8.1, 8.2
Characteristic Impedance	
Voltage Standing Wave Ratio	Objective 5.2
Digital RF Modulation	Objective 8.1
Amplitude Shift Keying	
Frequency Shift Keying	
Phase Shift Keying	
Spread Spectrum	

Chapter 2: RFID System Lifecycle	
Domain 2.0, 3.0, 6.0, 7.0	
Site Survey	Objective 6.1, 6.2, 7.1, 7.2
Installation	Objective 6.1
System Operation	Objective 7.3, Objective 5.2
Safety	
Electrostatic Discharge (ESD)	Objective 6.1, 2.2
Grounding and Power Supply Issues	Objective 6.1
Human Safety	Objective 3.4
Equipment Safety	
Chapter 3: Frequency Ranges	
Domain 5.0, 8.0	
Low Frequency Range	Objective 5.1
High Frequency Range	Objective 5.1
Ultrahigh Frequency Range	Objective 5.1
Microwave Frequency Range	Objective 5.1
Chapter 4: Antennas	
Domains 2.0,8.0	
Basic Operation	Objective 8.2
Antenna Radiation Patterns	Objective 5.2
Antenna Gain	Objective 8.3
Captured Power Density, Area, Power	Objective 8.3
Polarization, Beamwidth, and Bandwidth	Objective 8.3
Types of Antennas	Objective 5.2
UHF and Microwave Antennas	Objective 5.2.1, 4.1.2

Chapter 5: RFID Tags	
Domain 1.0, 4.0	
Tag Types	Objective 4.1
Tag Components	Objective 4.1
Passive Tags	Objective 4.1
Semi-Passive	Objective 4.1
Active	Objective 4.1
Classes	Objective 4.1
Class 0 Class 1	Objective 4.1
Class 1 Gen 2	Objective 4.1
Class 2	Objective 4.1
Class 3	Objective 4.1
Class 4	Objective 4.1
Tag Placement	Objective 1.3, 4.2
Chapter 6: Interrogators	
Domain 1.0, 2.0	
Interrogator Communication	Objective 1.1, 1.2
Interrogator Types	Objective 1.1
Handheld Interrogators	Objective 1.1
Mobile Mount Interrogator	Objective 1.1
Interrogator/Encoder	Objective 1.1
Software Configuration	Objective 1.1
Interrogation Zone	Objective 1.2, 2.1
Configuring the Interrogation Zone	Objective 1.2, 2.1

Dense Interrogator Mode	Objective 1.4

Chapter 7: RFID Regulations and Standards	
Domain 3.0	
The International Organization for Standardization (ISO)	Objective 3.1
Frequency Ranges	
ISO 18000 Air Interface Standards	Objective 3.2
Data Objects and Identifiers	
RFID in the Supply Chain	
Integrated Circuit Cards	
Agricultural Electronics	
Electronic Product Code (EPC)	Objective 3.1
The EPCglobal Network	
EPC Format	
EPC Standards	
International Standards	Objective 3.3
USA: Federal Communications Commission (FCC)	
Europe: ERO, CEPT, ETSI	
China: SAC	
Security	
Chapter 8: Applications	
Domain 9.0	
Part 1: Printers for RFID Labels	Objective 9.1, 9.2
Part 2: Smart Labels	
Part 3: RFID Applications	

Part 4: Improving Agribusiness	

For More Information

The best way to receive information about the exam is to contact CompTIA directly at their Web site, http://www.comptia.org

www.rfidnotes.com/RFID+

Briljent, LLC has developed a Web site for readers of RFID+ The Complete Review of Radio Frequency Identification. This site provides additional study resources and up-to-date information on CompTIA RFID+ certification. After purchasing this book, log on to the Web site using the following temporary username and password:

username: comptia exam
password: RFID06

Chapter 1
Primer

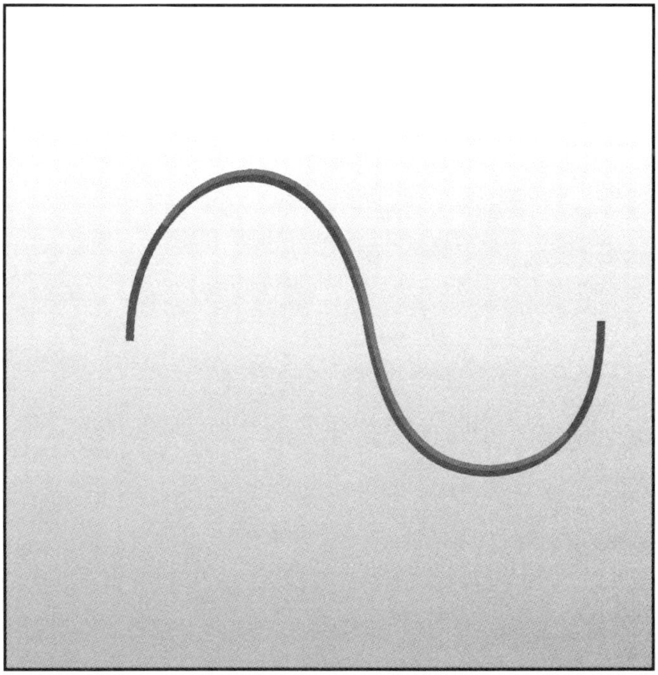

Chapter 1: Primer

Key Concepts

- Fundamental electrical concepts applied in RFID
- Faraday's Law for voltage induction in circuits
- Principles of magnetic fields
- Circuit coupling and RLC circuits
- The radio spectrum
- Frequency modulation

Terms

Amplitude modulation (AM)
Amplitude shift keying (ASK)
Attenuation coefficient
Backscatter
Characteristic impedance
Coupling
Diffraction
Direction
Faraday's Law
Flux line
Frequency
Frequency hopping
Frequency modulation (FM)
Frequency shift keying (FSK)
Inductance
Intensity
Keying
Magnetic fields
Modulation
Narrowband
Phase shift keying (PSK)
Propagation
Radiative coupling
Reflection
Refraction
Relative permittivity
Resonance
RLC circuit
Spread spectrum
Voltage Standing Wave Ratio (VSWR)
Wavelength

The focus of this chapter is to review the fundamental electrical concepts associated with RFID technology. A thorough review of these concepts provides a more comprehensive understanding of the RFID technology topics discussed throughout this book.

Faraday's Law

In the 1830s, physicist Michael Faraday first observed the phenomenon of voltage induction in circuits. The resulting physical law, **Faraday's Law**, is the principle that enables RFID systems to work. Faraday's Law states "voltage is induced in a circuit when the flux lines, which are proportional to the magnetic field, change in the circuit." Remember that **flux lines** are created by any current running through a wire.

Example

A coil of wire connected to a multimeter will create voltage if another magnetic field passes through it (Figure 1.1).

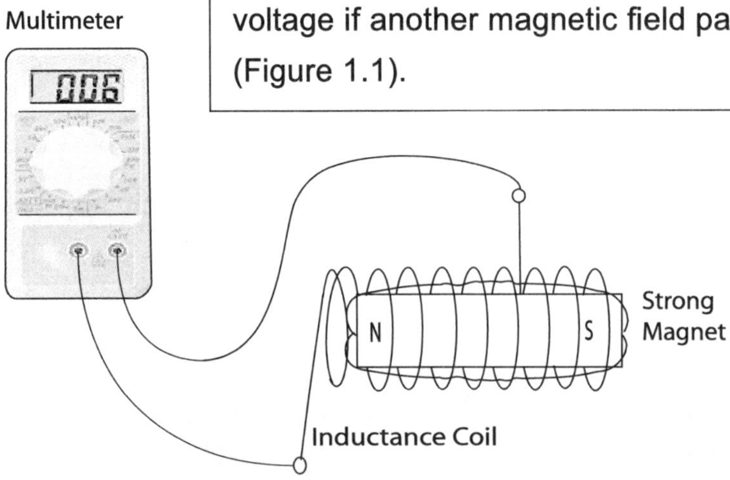

Figure 1.1 Voltage Created by a Multimeter

Magnetic Fields

Magnetic fields result from a charge moving through some form of conductor, e.g., current flow through a wire or a wire and a multimeter as shown in Figure 1.2. A magnetic field is a region in space in which a magnetic state exists (Fink 2004). While magnetic fields are invisible, they are definitely present. All electrical circuits have some degree of magnetic field.

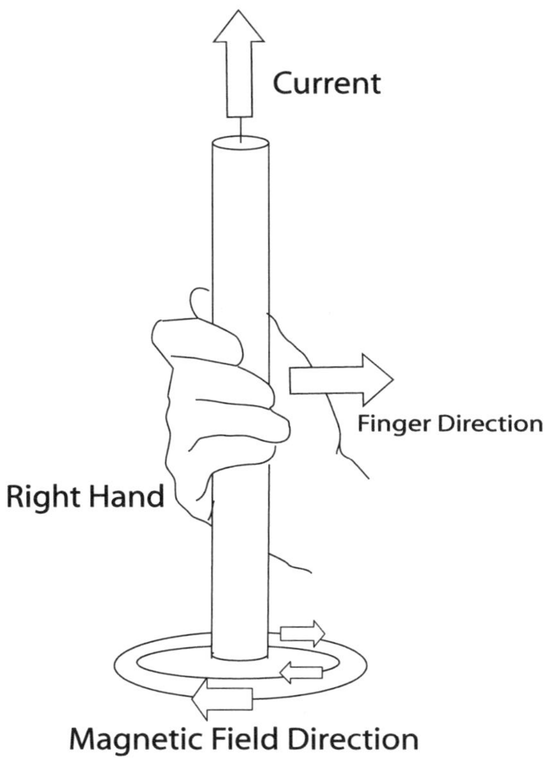

Figure 1.2 Right-Hand Rule of Magnetic Fields

There are two properties that exist at any point within the magnetic field region: **intensity** and **direction**. The lines of direction within a field, or flux lines, are commonly associated with the right-hand rule of magnetic force: If a person wraps his or her right hand around a conductor with the thumb pointing in the direction of the current, the index finger will point in the same direction as the flux lines around the conductor (Figure 1.2). Flux is not to be confused with magnetic fields or inductance. Flux lines merely indicate the direction associated with the field being produced. Accordingly, the magnetic field exists as a property of the circuit. **Inductance** is the circuit element that actually stores energy and creates the field.

Coupling

RFID systems operate using an electrical principle called **coupling**, which is the interaction or connection between circuits. Coupling occurs when the magnetic fields produced by one circuit overlap and link with another circuit's fields. There are different coupling systems (inductive coupling, capacitive coupling, and **radiative coupling** or **backscatter**) used for RF applications. Accuracy, size, cost, and range requirements help determine which type of coupling an RFID system will be configured to use, but inductive coupling systems are currently the most common. Inductive coupling occurs when a magnetic field created by one circuit enters the magnetic field of a second circuit. Inductance is an electromotive property created from the change in a magnetic field. The strongest form of inductance comes in the form of closed loops. The more loops or turns in an inductor, the stronger the field force. Inductors are effective in circuits because they harness this magnetic energy (Figure 1.3).

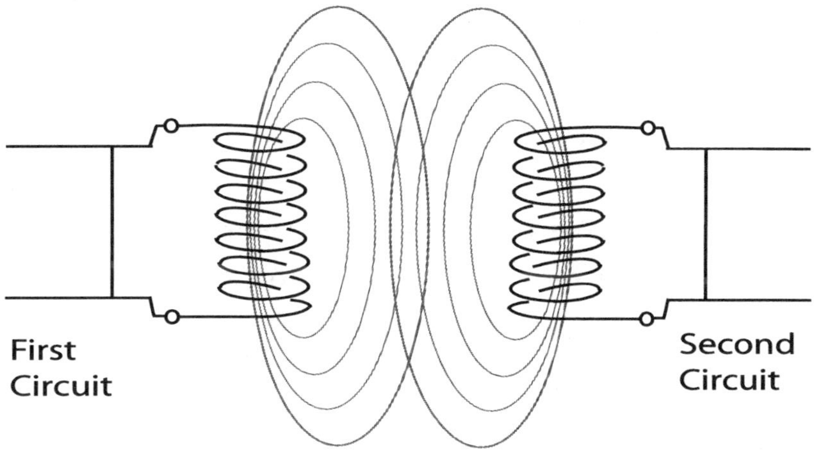

Figure 1.3 Inductive Coupling

Inductive coupling is how energy is transferred between the interrogator and tag for systems operations below (approximately) 27 MHz. At higher microwave frequencies, radiative coupling or backscatter transfers the energy. Backscatter is detailed in the discussion of tag and interrogator components.

An **RLC circuit** consists of three basic parts: the resistor (R), inductor (L), and capacitor (C). The capacitor stores electrical energy and presents this energy in the form of voltage. Similar to a capacitor, an inductor stores magnetic energy and presents that energy in the form of current. When combined in the same circuit, the inductor and capacitor will complement each other. The third element, resistance, is always present in any electric circuit. A resistor in an RLC circuit dampens the oscillations by removing energy from the circuit.

Resonance

To understand **resonance**, it should first be understood that circuits have a natural frequency. The combination of inductors and capacitors determines the natural frequency for a given circuit. To demonstrate this, first observe an LC circuit with no resistor present and an initial charge on the capacitor (Figure 1.4). At time equals zero (T = 0), energy is being stored in the capacitor, current is 0 (I = 0), and voltage is at a maximum.

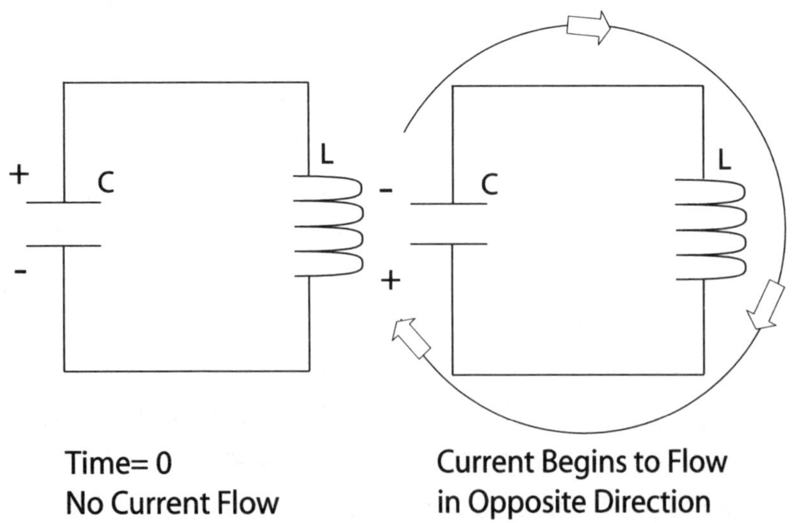

Figure 1.4 Oscillation Begins

1. The energy stored in the capacitor first flows from the capacitor into the inductor, leaving the capacitor with zero voltage and the inductor with all of the energy (Figure 1.5).

2. Meanwhile, the inductor generates a field that holds that energy from the capacitor.

3. Once all the energy is transferred from the capacitor to the inductor, the field in the inductor begins to collapse, and the energy flows out of the inductor back to the capacitor.

4. As the energy moves from the inductor to the capacitor, it charges up the capacitor with voltage, and the circuit condition is back ot its starting point (Figure 1.6).

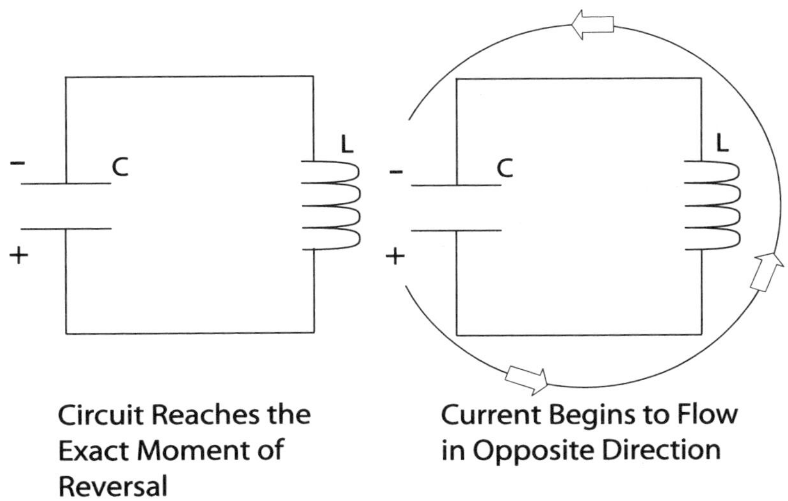

Figure 1.5 Resonance

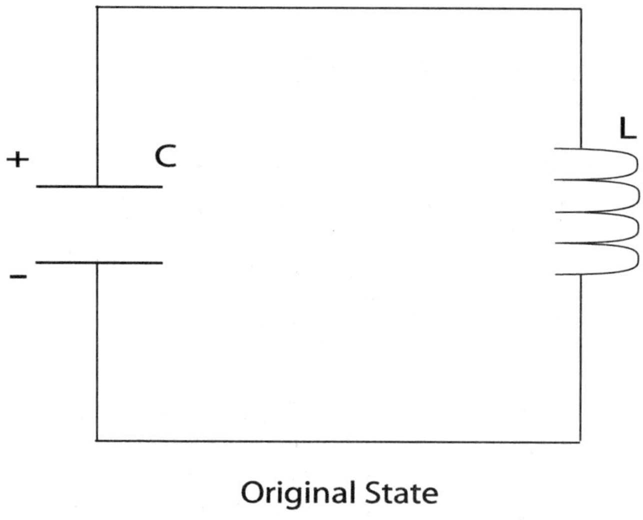

Figure 1.6 Circuit Starting Point

This repetitive transfer of energy, or oscillation, occurs at the circuit's natural frequency. The rate of this oscillation is dependent on the size of the inductor and capacitor and is defined as:

$$(1-1) \qquad f = \frac{1}{2\pi\sqrt{LC}}$$

The cycle could theoretically continue to repeat without end were it not for the inevitable existence of resistance within the circuit. Resistance in real-world circuits is impossible to eliminate. This resistance damps the amplitude of the current with each cycle until it is completely reduced to zero (I = 0). This damping can be overcome by adding an AC power supply to the circuit (Figure 1.7). A resistor is added to a circuit to manage the amplitude of current. The resistance of the circuit must be large enough to protect the circuit, but not so large that the current will not oscillate.

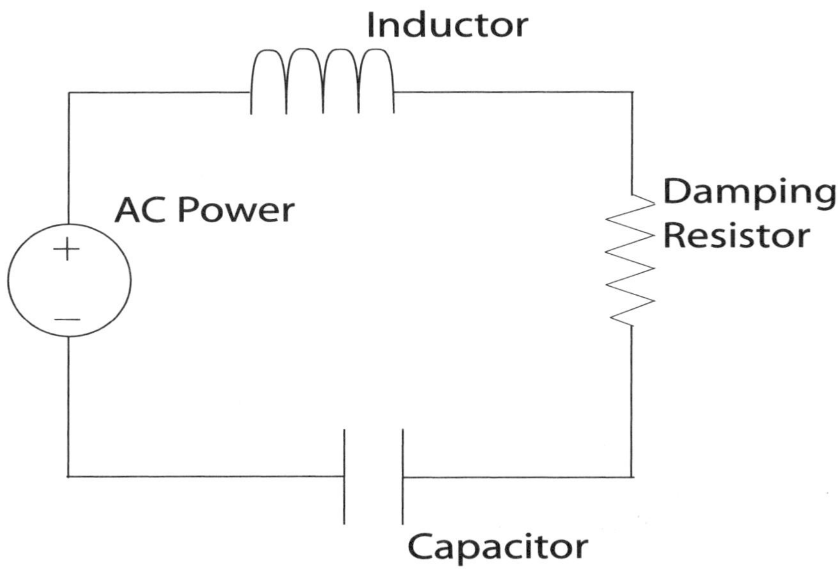

Figure 1.7 LC with Additional Resistance

In a parallel inductive-capacitive circuit (also known as a tank circuit) at resonance, the impedances are equal to each other (Figure 1.8). As a result, total impedance increases to infinity, and the tank draws no current from the power supply. In other words, at resonance the tank circuit acts as an open circuit. The frequency at which resonance occurs is the natural frequency, and it is strictly governed by the sizes of the inductor and capacitor. To calculate the resonant frequency of a parallel (tank) circuit with no resistance use the following formula:

$$(1-2) \qquad f_{res} = \frac{1}{2\pi\sqrt{LC}}$$

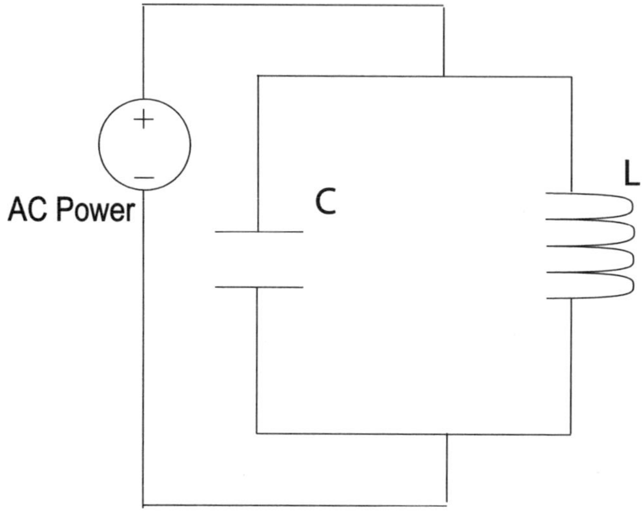

Figure 1.8 Parallel Inductive /Capacitive Circuit

In a series inductive/capacitive circuit, the impedances of the inductor and capacitor cancel each other out (Figure 1.9). This is the opposite of what occurs in a parallel circuit. As a result, total impedance drops to zero at resonance. In other words, at resonance in a series circuit, the result is a short circuit across the power supply.

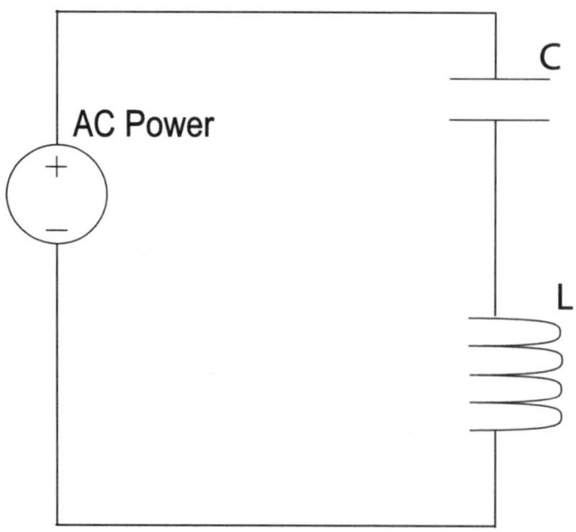

Figure 1.9 Series Inductive/Capacitive Circuit

Calculating the resonant frequency of a series circuit with no resistance is the same as with a parallel circuit:

$$(1-3) \qquad f_{res} = \frac{1}{2\pi\sqrt{LC}}$$

In summary, if the frequency of the AC power supply matches the natural frequency of the RLC circuit, resonance occurs. This frequency is called the resonant frequency. A distinctive characteristic of a resonant circuit is its bell-shaped response curve (Figure 1.10).

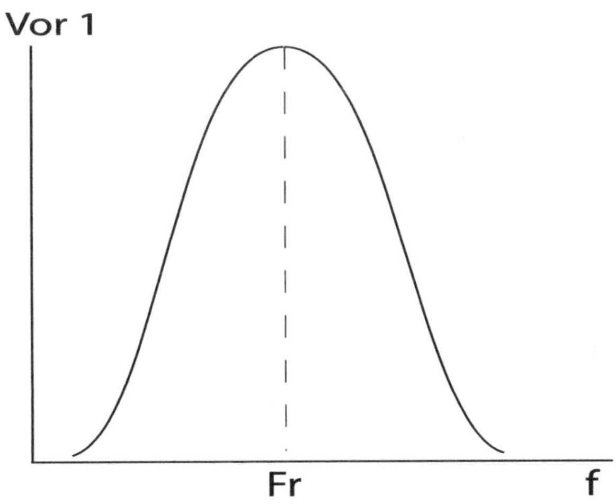

Figure 1.10 Frequency Response of a Resonant Circuit

Resonant circuits are the foundation of all radio frequency (RF) systems. The antenna must resonate at the frequency of the RF signal in order to transfer the maximum power. Additionally, the interrogator's amplifier must be tuned to the transmission frequency by adjusting the inductance and capacitance internal to the circuit.

RF Concepts

The RF spectrum spans from about 3 kHz to 300,000 MHz, encompassing the operation of everything from baby monitors to radio stations. Within this spectrum, RFID systems use four specific frequency bands:

1. Low Frequency (LF)
2. High Frequency (HF)
3. Ultrahigh Frequency (UHF)
4. Microwave

Frequency is a measure of the number of times a cycle repeats in a given period of time. Frequency is measured in hertz (Hz), which is equal to one cycle per second. Another method of evaluating radio waves is to use wavelengths. A **wavelength** is the length that one cycle of an electromagnetic wave occupies in space. Wavelength is proportional to the velocity of propagation and is measured in distance such as meters or inches. Wavelength, velocity, and frequency have a mathematical relationship that is expressed as:

$$(1-4) \qquad \text{wavelength} = \frac{\text{velocity}}{\text{frequency}} \quad \text{or} \quad \lambda = \frac{c}{f}$$

Electromagnetic waves travel at a constant rate of 3 x 108 m/s in free space. Because of this constant, there is a relationship between wavelength and frequency—higher frequencies have shorter wavelengths and lower frequencies have longer wavelengths. Period is the time it takes for a wavelength to go through one cycle. Period, denoted as T, is the inverse of frequency. That is:

$$(1-5) \qquad Z = \sqrt{\frac{\mu}{\varepsilon}}$$

Where Z_0 = character

Electromagnetic waves are essentially energy that is radiated through space. **Propagation** is the process of this electromagnetic energy traveling from the transmitter to the receiver. An electromagnetic wave is characterized by its method of propagation through an environment. When visualizing this phenomenon, it is common to picture energy traveling along a line-of-sight path, but line-of-sight is not the only model used.

Within the earth's atmosphere there are three methods of propagation: surface wave, sky wave, and space wave propagation (Figure 1.11). Consider how the earth's atmosphere is divided. The troposphere extends from the earth's surface to a height of 3.7 miles from the North Pole. The ionosphere extends from about 31.1 miles to nearly 250 miles above the earth's surface. The stratosphere is located between these two. Surface waves travel near the earth's surface within the troposphere by utilizing the earth's electromagnetic charge, and they follow the contours of the earth by means of diffraction. Sky waves are directed toward the ionosphere and are reflected or refracted back to the earth by the ionosphere. Space waves include both line-of-sight and ground-reflected waves and are normally found within the troposphere and stratosphere.

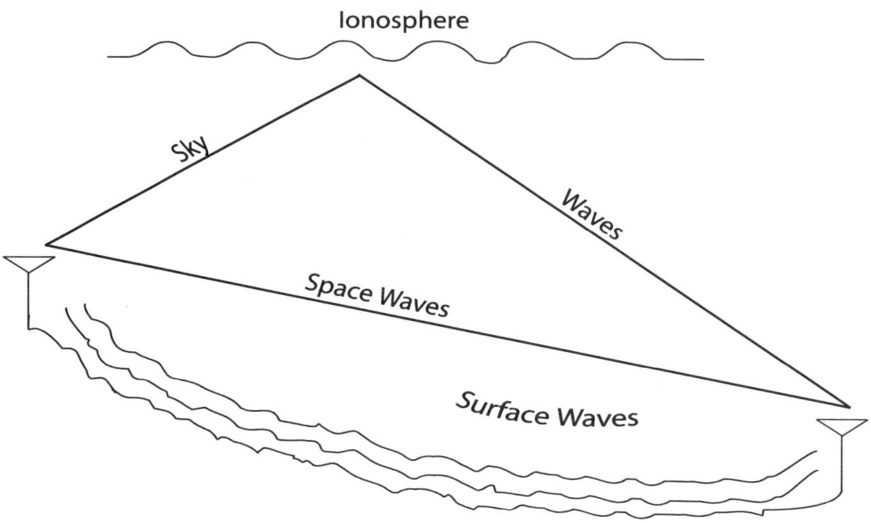

Figure 1.11 Space Wave Propagation

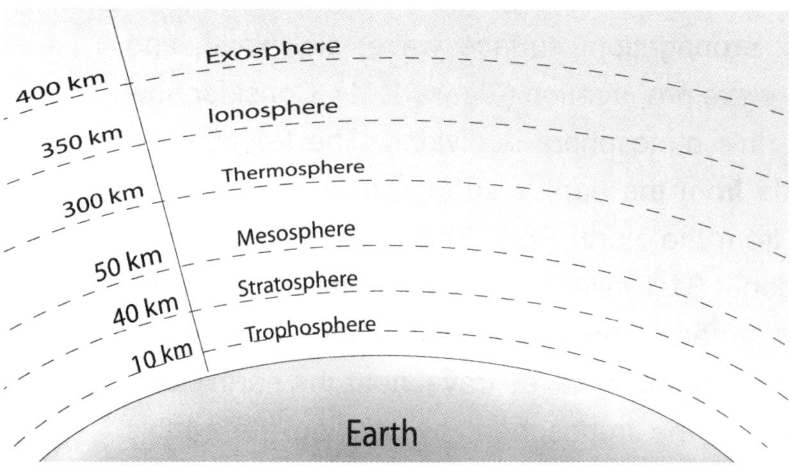

Figure 1.12 Space Wave Propagation

How Different Materials Affect RF

A wave will undergo different behaviors when it encounters an object in space. These behaviors allow RF signals to reach areas that cannot be reached by line-of-sight propagation. A few of these behaviors are reflection, diffraction, and refraction (Figure 1.13). **Reflection** occurs when a wave encounters an object and, rather than passing straight through it, the energy bounces off it in a new direction. **Diffraction** occurs when the radio path between the transmitter and receiver is obstructed by a surface that has sharp irregularities or edges. The secondary waves resulting from the obstructing surface are present throughout the space and even behind the obstacle, giving rise to a bending of waves around the obstacle. **Refraction** occurs when waves actually go through a substance, but not in a straight line. Waves are diffracted when they enter the object, which results in a change of the waves angle of direction. With light, this makes an object appear to bend at an air/water interface, as in Figure 1.12.

Signals behave according to the medium or substance in which they propagate. Substances are divided into three general types: conductors, insulators (also called dielectrics), and semiconductors. Generally, metals are conductors, nonmetals are insulators, and semiconductors are neither good conductors nor insulators. Semiconductors can act as insulators, but they can conduct only under specific conditions. When a radio wave comes in contact with a material, depending on this materials composition, some of the wave's power is absorbed and some of it is reflected. The power that is absorbed by the material is converted into heat. Conductors will reflect nearly all of the power, while insulators will only reflect a small amount of power. The capacity of a material to absorb or reduce power is its **attenuation coefficient**. Insulators have a low attenuation coefficient because, although some of power is absorbed, insulators allow most of the power to pass though. Conversely, conductors reflect most of the power, and thus have high attenuation coefficient. So, although insulators will allow RF signals to pass through, the speed at which the signal travels will be determines by the insulator's relative permittivity.

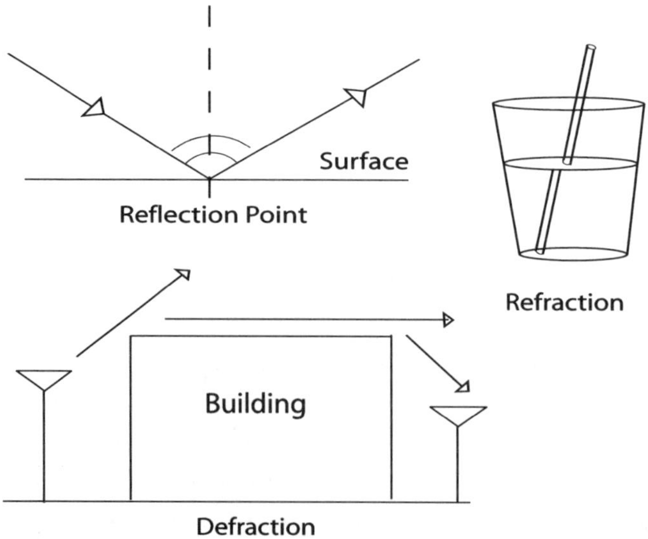

Figure 1.13 Space Wave Propagation

On an atomic level, conductors have many free electrons that are easily detached and move from atom to atom. The opposite is true for insulators, wherein electrons are so tightly bound that they are not often removed from their atoms. The **relative permittivity** (also called the dielectric constant) of an insulator is used to measure how easily it establishes electric flux in a material. The relative permittivity describes how electromagnetic signals move through the material. A material with a high relative permittivity will cause the signal to move more slowly than a material with a low relative permittivity. This makes the material look as if it is absorbing the RF signal. Table 1.1 shows the relative permittivity of some dielectric materials.

Table 1.1 Relative Permittivity of Dielectric Materials

Medium	Relative Permittivity
Vacuum	1
Air	1.0006
Styrofoam	1.03
Plywood	2.1
Polystyrene	2.7
Amber	3.0
Rubber	3.0
Plexiglass	3.4
Nylon	3.8
Sulfur	4.0
Quartz	5.0
Formica	6.0
Lead Glass	6.0
Mica	6.0
Marble	8.0
Ammonia	22
Glycerin	50
Distilled water	81

RF Transmission and Reception

A radio wave is used to carry information from one point to another through some form of medium, which can be wireless systems (Wi-Fi) or wired systems (copper cables).

In a typical RF system, RF waves carry information through the air to a destination point. The RF acts as a carrier that contains sidebands of data (Figure 1.14). This carrier frequency can be modulated in accordance with the data being sent. Once the signal is processed through the transmission circuitry, it is sent into the air through an antenna. The signal propagates through space, bouncing or deflecting off any material in the transmission area until it makes contact with a reception antenna. At this point, a signal goes through a receiver configuration that system demodulates the initial carrier from the information. The demodulation step allows the receiving device to manipulate output, or to respond to the initial transmission. Demodulation extrapolates the initial information that was modulated (Figure 1.15).

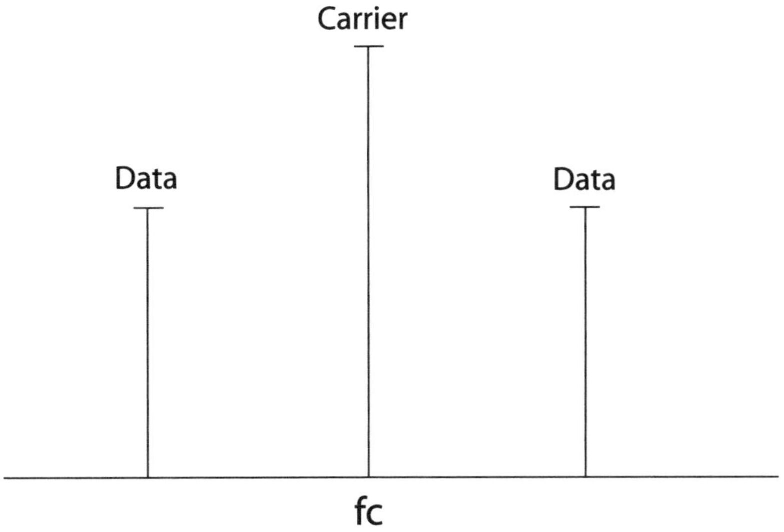

Figure 1.14 Transmission in the Frequency Domain

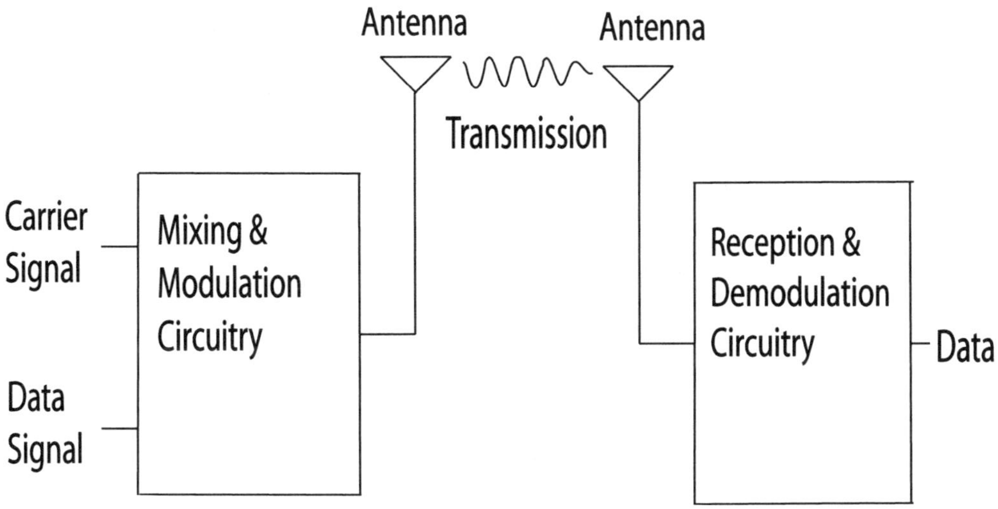

Figure 1.15 Demodulation

Characteristic Impedance

Now that the overall layout and operation of an RF system has been reviewed, **characteristic impedance** and the **Voltage Standing Wave Ratio (VSWR)** will be discussed.

Characteristic impedance, Z_0, is a figurative value of impedance for a transmission line that is infinitely long and lossless. The following is the equation for characteristic impedance:

$$(1-6) \qquad Z = \sqrt{\frac{\mu}{\varepsilon}}$$

Where Z_0 = character impedance
μ = magenetic permeability of the medium
ε = electric permeability of the medium

(The most obvious transmission line in an RFID system is the antenna cable.) Typically, character impedance is calculated for the lossless transmission line, and a resistance of that same impedance value is used to terminate the line. The matching of impedances makes the line look infinite. Transmission lines that terminate with their specific characteristic impedances produce no standing wave reflections. Standing waves are stationary as a result of two signals traveling in opposite directions adding together in the same transmission line. When there are no standing waves produced, the system achieves maximum power transfer between the transmitter and receiver as well as a constant ratio between voltage and current at any frequency and every point of the line. The characteristic impedance is the determining factor for how any RF transmission line will transfer power and operate. Impedance matching allows a system to maintain both the maximum power transfer and system stability.

Voltage Standing Wave Ratio

The power transmitted versus the power received along the transmission line is represented in the Voltage Standing Wave Ratio (VSWR), which is a measurement used to determine how well the components of an RF system are matched. The VSWR represents how much power is received at the end of the transmission line, which can be any medium between a transmitter and receiver. Consider the following scenarios to understand impedance and VSWR:

Scenario 1:

Suppose there are a source, a transmission line, and a load all matched to 50 ohms (Figure 1.16). If an RF system is properly matched at 50 ohms, the VSWR equals 1 and there is no RF power reflected back to the source. No voltage standing waves are produced, meaning that the maximum power transfer between RF components is achieved and there is no opposition between waves in the transmission line.

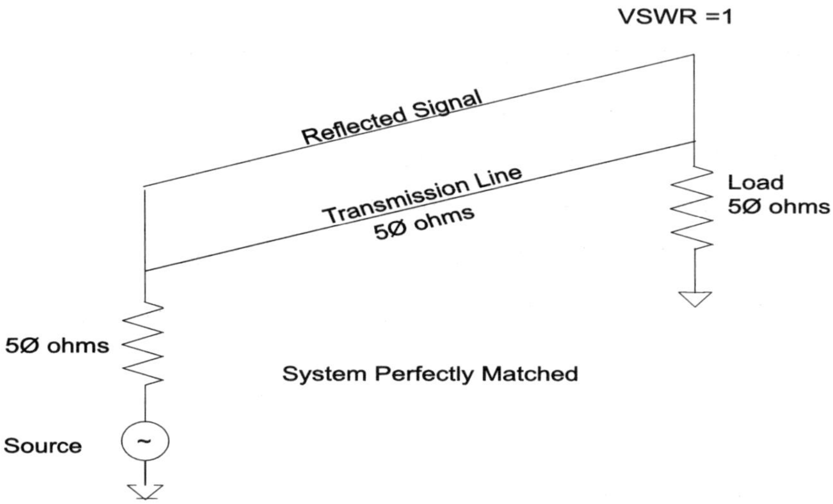

Figure 1.16 Scenario 1

Scenario 2:

Now, suppose the source and transmission line are matched to 50 ohms, but the load is 200 ohms (Figure 1.17). This system is not properly matched. The result is an undesirable standing wave produced by the reflection of the original signal once it reaches the receiver. The amplitude of the reflected waveform is far above our initial transmission. Most systems are not designed to take this kind of power reflection.

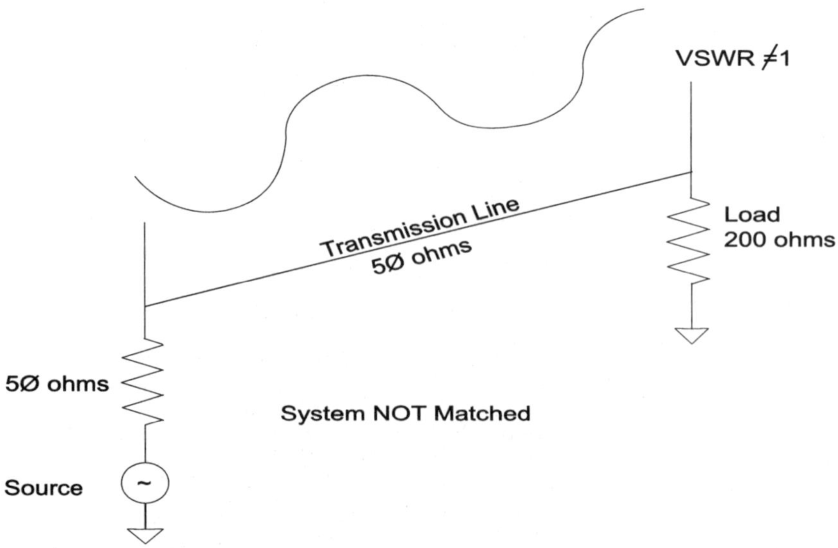

Figure 1.17 Scenario 2

Since no system can be perfectly matched, all transmission lines will have a VSWR other than one. Because of this, the length of the transmission line determines the amount of power that will be delivered to the load. If the line is terminated at a point where voltage is at a peak, the load will receive a stronger signal than if the load is situated at a voltage minimum.

It is important to consider the characteristic impedance and the VSWR because an RFID system is directly affected by these characteristics. The antenna and cable length are designed by the manufacturer to work with a specific interrogator according to FCC antenna output power standards. Any adjustment made to cable, antenna, or interrogator can cause the VSWR and the characteristic impedance to change in the RFID system. In turn, the output power of the antenna could change, causing system instability. Moreover, such a change may violate regulations or standards.

Modulation

Modulation is the alteration of a high-frequency carrier as determined by a baseband signal—a simple concept with broad connotations. RF modulation is the process of combining information or baseband data with a high-frequency carrier for transmission from point A to point B through a medium. Modulating a signal involves altering the high-frequency carrier signal's frequency, phase, or amplitude. The data signal is mixed with the carrier to produce a modulated signal to be transmitted. As a data signal comes into the system, it is mixed with the carrier to produce a modulated signal to be transmitted (Figure 1.18).

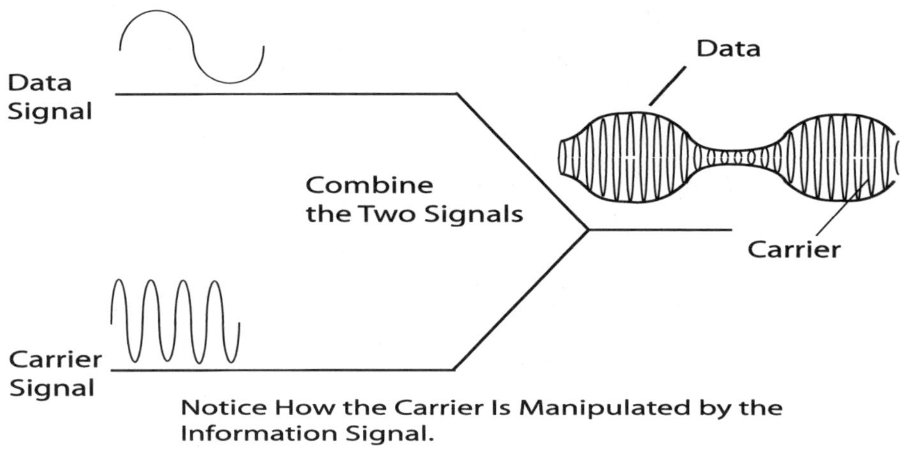

Figure 1.18 Basic Modulation

It is important to note that all forms of modulation, by definition, create additional sidebands of data. These sidebands are a result of the actual modulation process. The definition of a sideband is a band of frequencies located either above (upper sideband) or below (lower sideband) the carrier frequency. Sidebands are more commonly referred to as the upper sidebands (USB) and the lower sidebands (LSB), which are the locations in the frequency domain of the actual data a system transmits (Figure 1.19).

Information is Located on the LSB and USB (Frequency Domain)

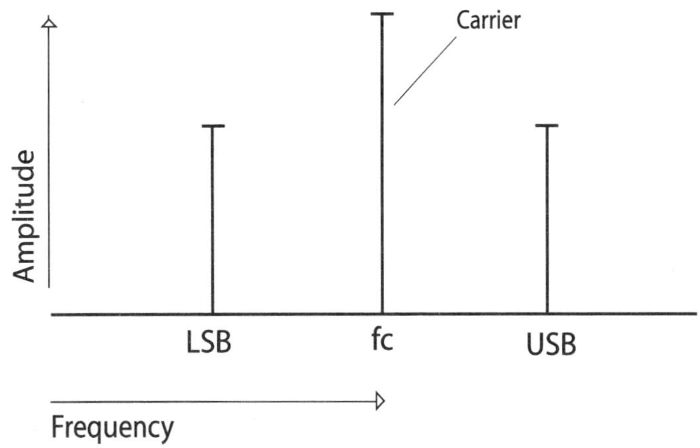

Figure 1.19 LSB and USB Sidebands

There are two basic types of RF modulation: **amplitude modulation (AM)** and **frequency modulation (FM)**. These two modulation types are the basis for radio wave communications and should be familiar terms. The more complex modulation schemes are derived from these two basic methods.

AM occurs when the amplitude (or voltage) of the carrier frequency changes with respect to the low-frequency modulated signal. The carrier frequency develops peaks and valleys according to the modulated signal (Figure 1.20). There are several types of AM, including double-sideband full carrier (DSB-FC), double-sideband suppressed carrier (DSB-SC), and single-sideband suppressed carrier (SSB-SC). AM is the most fundamental form of modulation used in RF transmission.

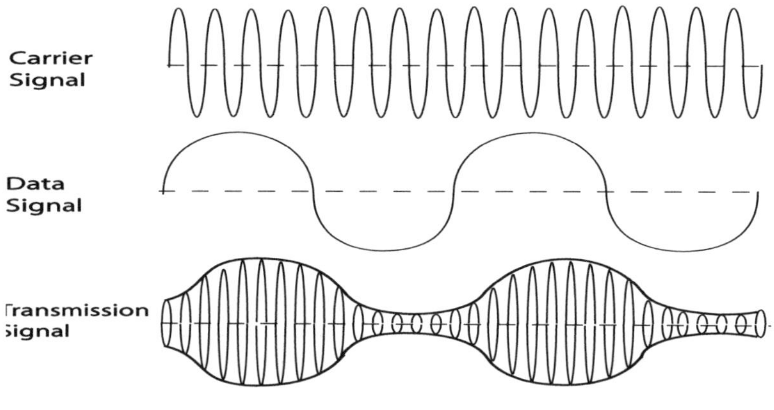

Figure 1.20 Amplitude Modulation Signal

FM occurs when a carrier signal's frequency is manipulated by baseband (low-frequency) data. Figure 1.21 shows how the carrier becomes compressed or expanded relative to the information signal. Notice that amplitude remains the same. When amplitude is static, noise has much of an effect on the signal. There is constant voltage, and the information is carried on a strong signal.

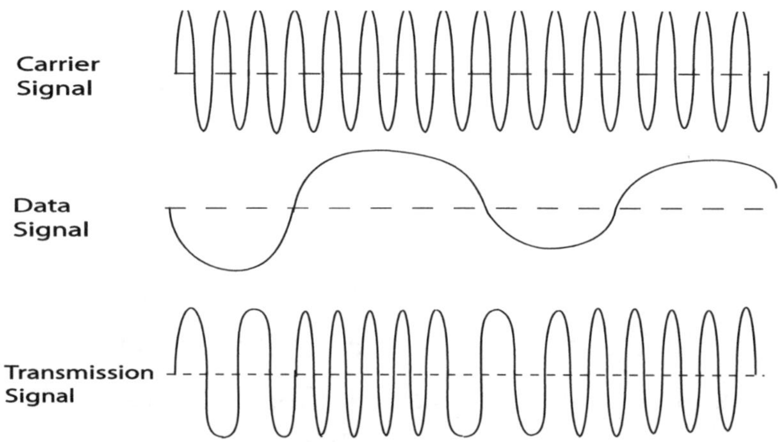

Figure 1.21 Frequency Modulation Signal

Digital RF Modulation

The concepts from this point forward cover a more complex form of RF modulation. It is important, though, to understand the basics before continuing because all forms of complex modulation stem from the basic modulation techniques.

Digital modulation connotes the presence of digital data associated with a system. Whenever digital data are involved, modulation includes a process known as **keying**. Keying means that one or more properties of the carrier signal are shifted between a set of states. Taking a brief look at each of the three types of basic modulation helps to explain the keying process.

Amplitude Shift Keying

Amplitude shift keying (ASK) is a more complex form of AM (Figure 1.22). ASK occurs when digital data are represented as variations in the amplitude of a carrier wave. The simplest form of this modulation technique is called switching or on-off keying. The amplitude of the carrier is switched between a binary 1 (on) and 0 (off). Note how this modulation technique operates. When the binary input of the system is a 1 (high), the amplitude and frequency of the carrier remain constant. When the binary input is a 0 (low), the carrier is turned completely off.

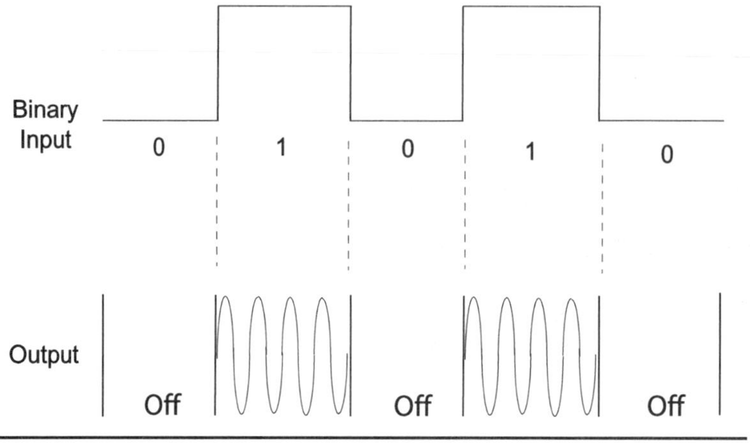

Figure 1.22 Amplitude Shift Keying (ASK)

Frequency Shift Keying

Frequency shift keying (FSK) occurs when the carrier is varied between two frequencies according to the binary inputs of the system. The amplitude is the parameter that remains constant in this modulation type. The two frequencies referred to are the mark and space frequencies. A binary 0 at the input of the system results in the space frequency being the output (Figure 1.23). A binary 1 at input results in the output of the mark frequency. The carrier is modulated according to these frequencies.

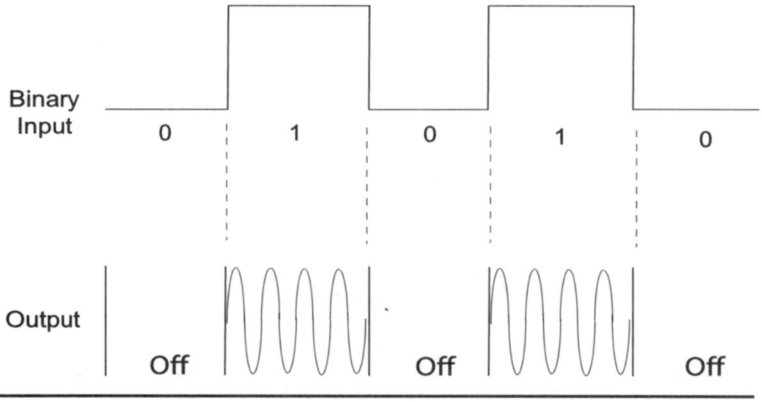

Figure 1.23 Frequency Shift Keying

Phase Shift Keying

Phase shift keying (PSK) is a form of digital modulation often used for the transmission of data across phone lines. There are a limited number of output phases available with PSK. The binary information used as the input data is encoded into bits before modulating the carrier. The binary 0 input causes a shift of 0 degrees in the carrier signal. The binary 1 input causes a shift of 180 degrees. Even though phases are changed in this modulation scheme, the carrier's amplitude remains constant (Figure 1.24). There are also other applicable PSK methods: quadrature phase shift keying (QPSK) involves four output phases that can manipulate a carrier signal using two bits; 8-PSK allows for eight output phases to alter the carrier signal using three bits; and 16-PSK gives the option of having sixteen output phases using four bits.

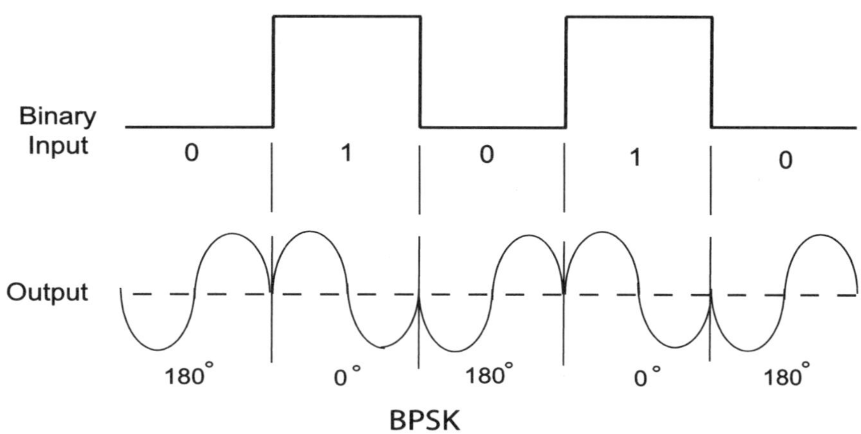

Figure 1.24 Phase Shift Keying

Spread Spectrum

Narrowband is used to describe a radio signal that occupies a small amount of space on the radio spectrum. The FCC tends to prefer narrowband signals, but a major drawback is that other signals at or near a narrowband signal's frequency can cause interference. **Spread spectrum** is an RF communications system in which the amount of RF spectrum used is much greater than the bandwidth of the baseband signal. A direct sequence system uses a carrier fixed to a specific frequency band, modulating it with a pseudo-random code. The resulting signal is spread over a much larger amount of bandwidth than the original baseband signal. The receiver uses the pseudo-random sequence and the demodulated spread signal to produce the original baseband signal.

The added capacity of the spread signal reduces the overall power of the transmitted signal. Another advantage of the direct sequence system is that it actually transmits at least ten redundant copies of the data at the same time. This helps reduce the impact of interference since only one copy needs to be received. Because the process requires the signal to be despread at the receiving end, any interference that is present will also undergo the de-spreading process. This causes a dramatic reduction in power density of the interference and greatly reduces or eliminates the interference. In comparison with a narrowband signal, a direct sequence signal has a lower power density that makes it more difficult to detect, thus making it more secure.

Frequency Hopping

Another type of spread spectrum technique is frequency hopping. Rather than modulate or spread the signal with code, the carrier frequency is charged at short intervals, making the signal hop around within a section of the frequency domain (Figure 1.25). As a result of the spreading and de-spreading of the signal when sending and receiving, the signal processing gain improves the signal to noise ratio (S/N ratio). Since the frequency hopping technique involves no spreading or de-spreading, it cannot achieve the same ratio as the direct sequence system. In addition, it is more difficult to synchronize the receiver and transmitter in a frequency hopping system than in a direct sequence because both time and frequency need to be in tune. This increases the latency time between data transmissions, considering frequency hopping radios must spend more time to search for the signal and connect to it; however, since the data streams between interrogators and tags are often very short, all of the transmission can be sent in one packet and this latency time is not an issue.

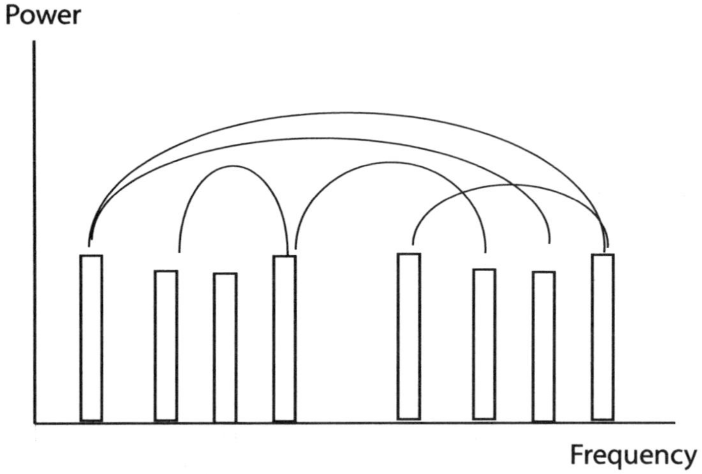

Figure 1.25 Frequency Hopping

Summary

Understanding the fundamental electrical principles behind RF is the first step in building a comprehensive skill set in RFID technology, which is built around the knowledge of how radio signals interact with each other and with their environment. The basic principles of circuits, electromagnetics, and radio frequencies affect the functionality of all RFID components and must be accounted for when planning, configuring, and maintaining a system.

Review

Questions

Answers are provided in Appendix A.

1. What units are used to measure inductance?
   a. Ohms
   b. Volt
   c. Watt
   d. Henries
2. Which will have a longer wavelength?
   a. signal of 10 MHz
   b. signal of 50 MHz
   c. signal of 100 MHz
   d. signal of 200 MHz
3. The correct relationship between period and frequency is?
   a. f = 1/T
   b. T = 1/F
   c. fT = 1
   d. all of the above
4. In an inductor, what is the cause of opposition to the circuit?
   a. reluctance
   b. resistance
   c. capacitance
   d. none of the above

5. What VSWR will result if a 50-ohm load is connected to an antenna with a 10-ohm feed point impedance?\
   a. 1
   b. 1.5
   c. .5
   d. .5
6. What is the resonant frequency of a series circuit made with a 10-mH inductor and a 3-uF capacitor?
7. An RF signal bounces off a metal frame. This is an example of _____.
   a. refraction
   b. diffraction
   c. reflection
   d. sky waves
8. An RF signal bends around a wall. This is an example of _____.
   a. refraction
   b. diffraction
   c. reflection
   d. ground waves
9. When RF energy is absorbed by an object, it is usually transformed to ____.
   a. light energy
   b. induction
   c. magnetic field
   d. heat energy

10. Which has a higher attenuation constant: insulators or conductors?

11. An RF signal travels faster through _____.

    a. air
    b. water
    c. glass
    d. rubber

12. True or False: All the power in an RF signal is contained in the carrier.

13. True of False: Only AM modulation creates side bands that carry information.

14. Spread spectrum systems, when compared to narrowband systems _____.

    a. use more bandwidth
    b. use less power
    c. are less prone to interference
    d. all of the above
    e. both a and c

15. Amplitude Shift Keying is an example of _____.

    a. amplitude modulation
    b. digital modulation
    c. spread spectrum modulation
    d. both a & b

Discussion

1. Why is it important that the impedances of an RFID antenna, interrogator, and cable be matched?

2. Put the following materials in order from fastest to slowest signal propagation (based on the relative permittivity).

   - Rubber
   - Mica
   - Plexiglas
   - Air
   - Glycerin
   - Nylon

3. What effect can the length of an antenna cable have on the operation of an RFID system? What causes this effect? What effect will a material with high relative permittivity have on an RF signal?

4. Define the following terms:

   - Inductance
   - Coupling
   - Capacitance

Chapter 2
System Lifecycle

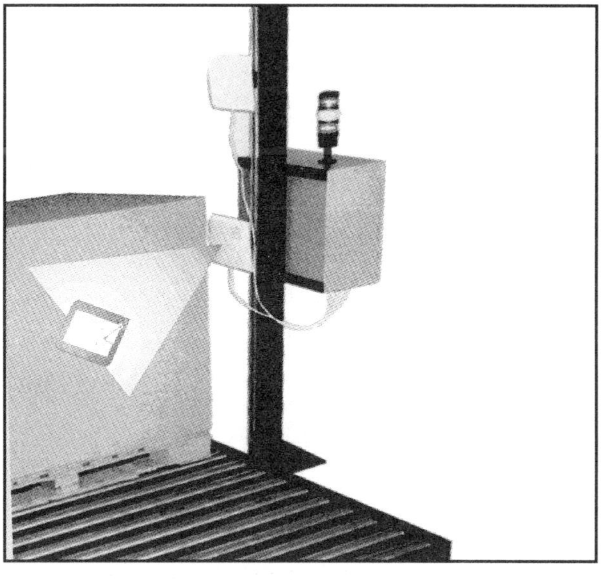

Chapter 2: System Lifecycle

Key Concepts

- Site survey—purpose and procedure
- Installation procedures
- System component
- Safety of people and equipment

Terms

Antennas

Earth ground

Electrostatic Discharge (ESD)

Full Faraday Cycle Analysis

Ground loop

Host computer

Interrogator

Interrogation zone

Medical implant

Radiation patterns

Radio frequency modulation

Spectrum analyzer

Tag

Voltage Standing Wave Ratio (VSWR)

Before discussing the intricacies of each component of an RFID system, it is important to consider the larger picture, from the site survey to installation and maintenance. Keeping the overall lifecycle of a system in mind while examining the details will give the reader a better vision of how all of the parts are connected in the system as a whole.

Site Survey

The purpose of a site survey is to identify sources of RF interference, with the goal of facilitating RFID system operation. The signal coming from the tag to the interrogator is very small, about -80 dB, compared to the signal from the interrogator. If other RF sources are too strong, they can swamp the signal coming from the tag and confuse the interrogator. This is a problem with a low signal-to-noise ratio: the signal from the tag is too small to override the noise coming from other RF sources. If it is not known that other RF sources are preventing tags from being read, hours can be spent troubleshooting an RFID system that would work perfectly in a “quiet” environment. A site survey should be done before completing the final design and installation of an RFID interrogator or antenna.

The requirements and complexity of a site survey vary depending on the facility. For any installation, it is crucial to know that there will not be interference inside the interrogation zone and that the interrogators will not interfere with existing RF systems. The focus of the most thorough measurements will be at the projected RFID interrogation zones, but all potential sources of interference must be located.

The first step of a site survey is to obtain a facility diagram or blueline drawing. If neither is available, a sketch to scale may have to suffice. A visual inspection of the facility should be done to confirm the accuracy of the prints. Wireless LAN access points, cordless phones, 2-way radios, and other RF sources, should then be marked.

The next step is to identify where the RFID interrogation zones are most likely to be installed. A **Full Faraday Cycle Analysis** is performed to identify all possible sources of interference at all possible times of operation. This usual requires 24–48 hours of data, or at the minimum, a full business day that observes all shifts. Data representing only a snapshot of one time of day is inadequate. Many sources of interference, such as handheld radios or cleaning equipment, are intermittent. If it is possible for the intermittent equipment to be on during this time frame, field strength data should be recorded.

The tool used to measure field strength at specific frequencies is a **spectrum analyzer**. Whereas oscilloscopes measure voltage versus time, spectrum analyzers measure voltage (or power) versus frequency (Figure 2.1).

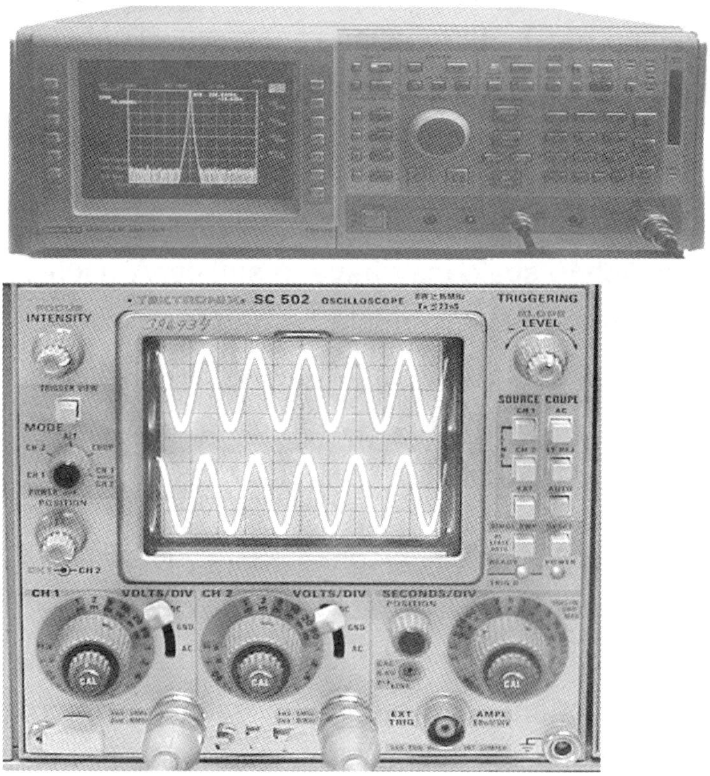

Figure 2.1 Spectrum Analyzer and Oscilloscope

In the spectrum from an interrogator, the signal will hop from channel to channel in the frequency band allotted. Regulations prohibit an RFID signal resting at any one frequency for more than a fraction of a second. The spectrum analyzer must be able to cover the whole span of the RFID frequency hopping, plus a comparable frequency span above and below. For example, for a 915-MHz system, the interrogator's signal actually hops between several bands from 902 to 928 MHz. The spectrum analyzer should display a 60 MHz span from 885 to 945 MHz. Setting the analyzer's display for "high persistence" or "store" will capture any spikes that might occur while you are not monitoring the screen.

To capture the RF signals, an antenna will be connected to the input of the spectrum analyzer. The antenna should be mounted in the middle of proposed interrogation zone. A tripod may be used, or the antenna can be suspended using rope (or something else nonmetallic). The antenna must be tuned to a frequency of interest (such as a ½-wavelength) to take accurate readings. Information about the antenna's gain and impedance (sometimes given in an "antenna table") will allow translation from dB displayed on the spectrum analyzer to actual signal strength density (in power per square meter). Without this calibration information for the antenna, only relative power can be measured. While one may still find sources of interference and observe where the interference is the greatest, the exact field strength of the interference cannot be determined.

The easiest way to collect a day's worth of data is to use a spectrum analyzer that can interface with a laptop computer. Not all analyzers have this option, but some come with interfaces for RS232, Ethernet, or GPIB. Software must be programmed on the laptop to collect data from the analyzer on a regular schedule (at least once an hour, but more frequent collection is better for pinpointing intermittent sources of interference).

If time permits, various locations (such as the corners of the facility) should also be tested. After all the data are collected, they must be analyzed for RF signals in the frequency range where the RFID system will operate. If there are spikes recorded, the source of interference should be found to determine whether the source can be removed, or whether the RFID installation will have to be shielded from it. The antenna and spectrum analyzer may be used to triangulate the source of interference if it cannot be located visually. This may mean moving the test setup around the facility. All sources of interference should be labeled on the facility diagram. This will ensure the best possible RFID installation.

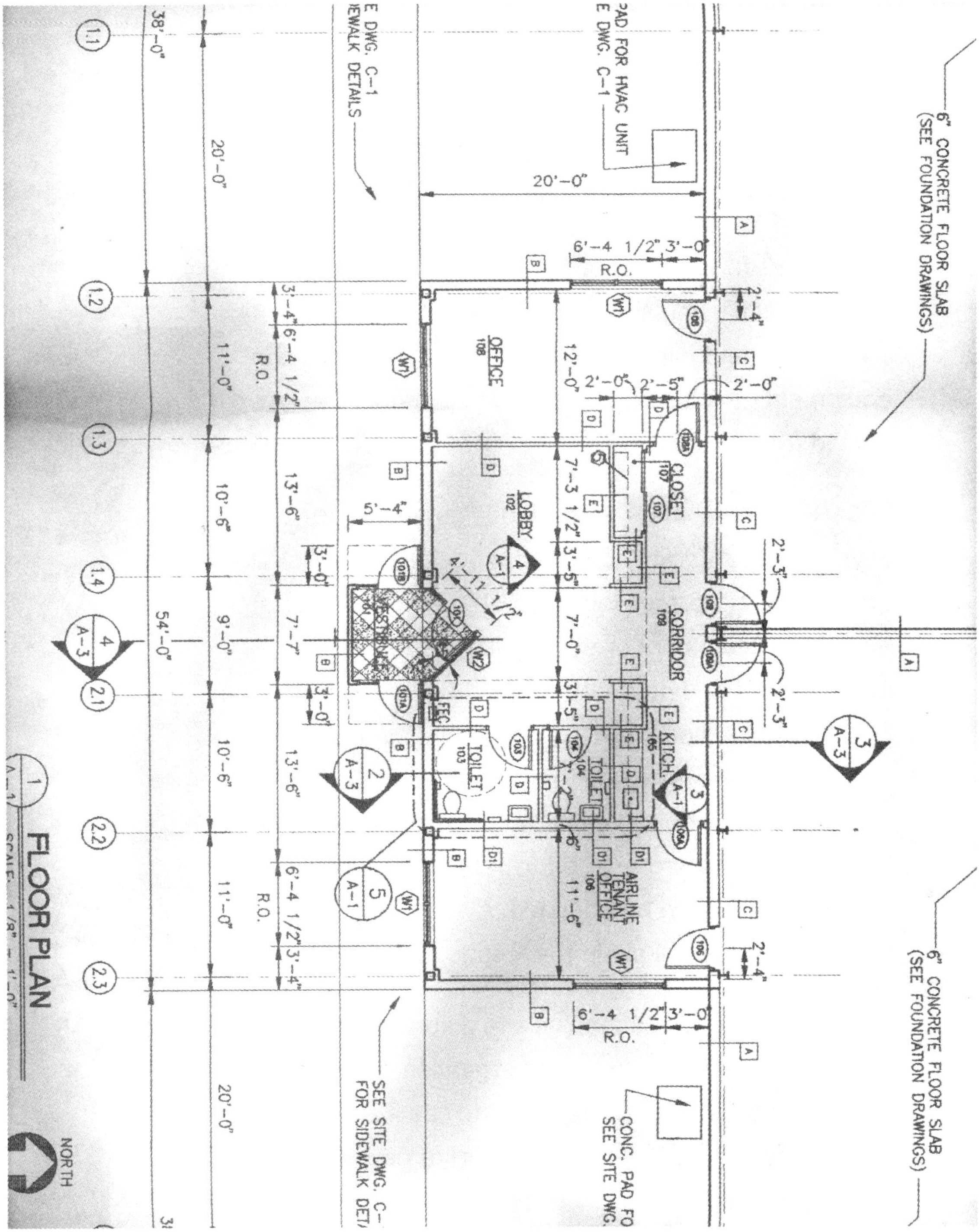

Figure 2.2 Blueline Drawing

Installation

After the site analysis has been conducted, the information will be given to an RFID system architect who will design the installation. The design process falls outside of the RFID+ certification exam domain. The design will then be communicated to the technologist, typically using site diagrams, blueline drawings, parts lists, and other documents that describe the equipment, its placement, and how it should be connected (Figure 2.2).

The RFID equipment should be tested before the installation, preferably before the equipment is delivered to the customer location. This will ensure that any faulty equipment or system integration problems are identified and corrected without tying up the customer's time or floor space. Additionally, all documentation should be reviewed to make sure it is complete and coherent. Copies of the latest firmware and any application software needed to test and validate the system should also be obtained. A supply of properly functioning tags will aid in the testing.

At the installation location, the site diagram should be used to locate the planned interrogation zones. The locations should then be observed to see if any potential RF interference sources (such as a new wireless hub) have been added since the site survey.

Mounting

Antennas are mounted according to the site diagram so that their **radiation patterns** cover the interrogation zone. Many installations require more than one antenna tied to the same interrogator. The site diagram will also show where the interrogators are mounted. All antennas and interrogators must be securely mounted with the appropriate hardware. The technologist must be sure that the hardware used to hold the equipment does not interfere with the RF signals. Metal brackets must not be used, since they will disrupt the antennas' radiation patterns. On the other hand, most plastics are transparent to RF signals and can be safely used.

Antennas may need additional enclosures to prevent damage from fork lifts and other moving objects. Any enclosures must not interfere with the radiation patterns. Sometimes metal may be used to support an enclosure, with the support's reflective characteristics being intentionally used to contain the interrogation zone. Using RF reflective or absorptive materials can also have negative effects on the interrogation zone. Any enclosure design must be tested before installation to assure the proper interrogation zone is established.

Cables

Proper cables (probably supplied with the antenna from the manufacturer) must be used to connect the interrogator to each antenna. A cable of the wrong length or configuration can impede operation or even damage equipment. If cables must be fabricated, the architect's instructions for length, type, and connectors must be followed. Cable locations (especially those connecting to the customer's network or power infrastructure) will be indicated on the site diagram. Cables are fragile and must be protected, and can also be a tripping hazard. This is why cables should be housed in conduit. A technologist should make sure there is adequate room in the conduit for pulling cables in and out so that no cables are pinched.

System Operation

Once everything is mounted, connected (including power), and verified according to the site diagram, the system can be powered up. Software versions should be verified, and then a system test can be performed. First, the interrogator's response to known good tags is tested, and then the read rates are tested using the customer's items. The direction of antenna's signal should checked to establish the desired zone configuration. The interrogator power can also be adjusted within limits to assure good interface with tags without interference with other RF systems or interrogation zones. When the system has been tested and appears to be working according to specifications, another site analysis should be performed to verify that RF energy from the new RFID system is contained within the interrogation zone and is not interfering with other systems.

The Overall Systems

An RFID system consists of an **interrogator** (or reader), **tags**, tag and interrogator **antennas**, and a **host computer** (Figure 2.3). For the purposes of this book and RFID certification, host computers and middleware will only be covered in a broad sense. Host computers and middleware operate in myriad and model-specific ways, so the precise mechanics and functions of these types of hardware fall outside a technologist's concern.

When an interrogator first sends an RF signal to an antenna, RF modulation occurs (Figure 2.4). In **RF modulation**, the initial signal is encoded and prepared for transmission. If the antenna has been connected to the interrogator with the correct length of cable (considered a transmission line), the signal will be transmitted without distortion. Output power is also established at this step in the process. For this reason, the configuration of the interrogator and antenna determine the output power of a system.

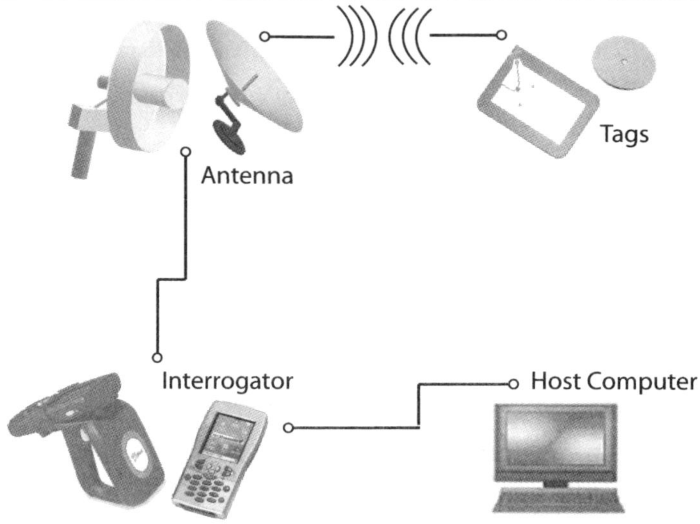

Figure 2.3 Overall System Operation

When the interrogator antenna transmits the RF modulated signal, the signal travels the farthest distance it can, according to the antenna's radiation pattern (Figure 2.5). The **interrogation zone** is defined during the antenna's transmission. Before operation, the initial setup and testing of the system will have established the range and integrity of the propagated signal. If the components have been set up perfectly and the system has been properly analyzed, tags in the interrogation zone will be read 100 percent of the time.

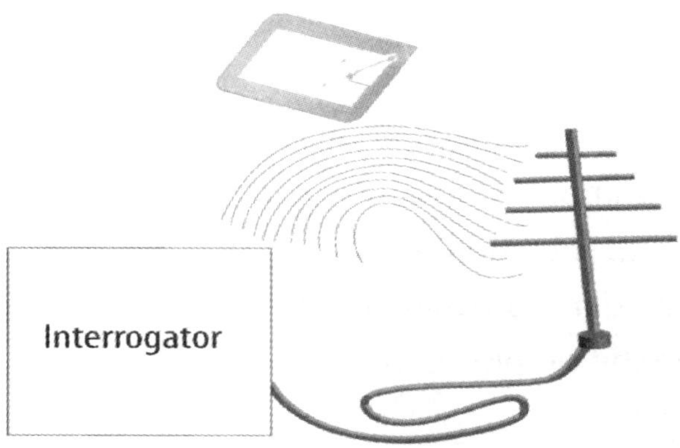

Figure 2.4 Antenna to Tag

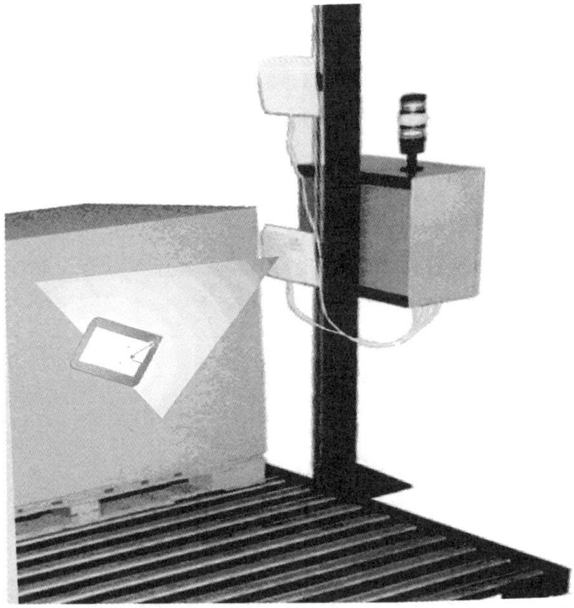

Figure 2.5 Antenna Propagation

When a tag receives power from the initial carrier signal, it is activated or "wakes up" and starts operation. By this time, inductive coupling or backscatter will have occurred. The tag continues using the interrogator's signal to gain power until it can generate a return signal. Tags determine how much a system will be able to accomplish because they retain all of the pertinent data a system is working to transport. Once the return signal is generated, the tag transmits it to the antenna, which receives the tag's information and communicates it to the interrogator (Figure 2.6). As when the signal is first sent from the interrogator to the antenna, it is important that the proper connections be maintained with the **voltage standing wave ratio (VSWR)** as close to one as possible.

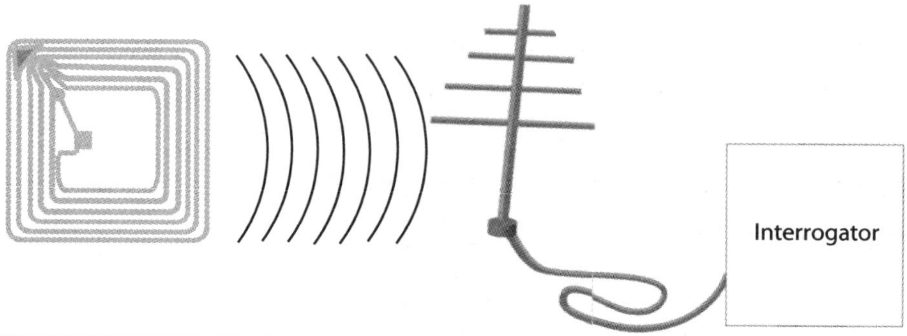

Figure 2.6 Tag to Antenna

After receiving tag data from the antenna, the interrogator communicates this information to the appropriate middleware. From the middleware, the flow of data is organized so that a company's database or host computer can manipulate the information for myriad uses.

Safety

Electrostatic Discharge (ESD)

In any manufacturing or shipping/receiving environment there are many sources of **electrostatic discharge (ESD)** that can damage equipment. Any place two surfaces rub or separate repeatedly is a potential source for thousands of volts of static electricity. Common sources of ESD include conveyor belts, rollers, stripping labels from rolls, and paper handling. The high potential created by ESD can find its way across tags and burn out the silicon transistors at the input or output of the tag's IC, making the tag inoperable. ESD damage to tags is a common cause of tag malfunction.

ESD damage to the interrogator is less likely, but still possible. The interrogator RF inputs are highly sensitive to the small signals coming from the tags. If an interrogator is not properly grounded, the ICs inside the interrogator may be damaged by ESD.

Grounding and Power Supply Issues

In order for any electrical system to work safely and reliably, it must have a reliable source of power with adequate grounding. Most permanent facilities have an **earth ground** with the AC power line, providing a safe path to ground for any AC power short that might occur, as well as a path for static electricity to discharge. In temporary field locations (perhaps using a generator for power) a ground rod will have to be deployed to supply an earth ground (Figure 2.7). Lack of proper grounding is not only dangerous to humans, but can also damage equipment when ESD builds up and discharges through components.

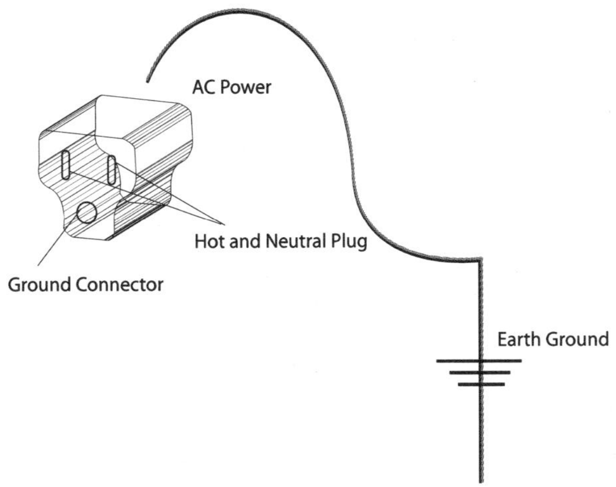

Figure 2.7 Earth Ground

Ground loops occur when current travels along ground wires in a circular path between components (Figure 2.8). This configuration must be avoided by making sure there is not a loop when the equipment is wired. The preferred configuration is a common ground (Figure 2.9) in which all ground wires meet at one point. Ground loops create a potential at the component ground node and prevent proper operation. This might result in intermittent operation or may prevent any operation, depending on the sensitivity of the equipment.

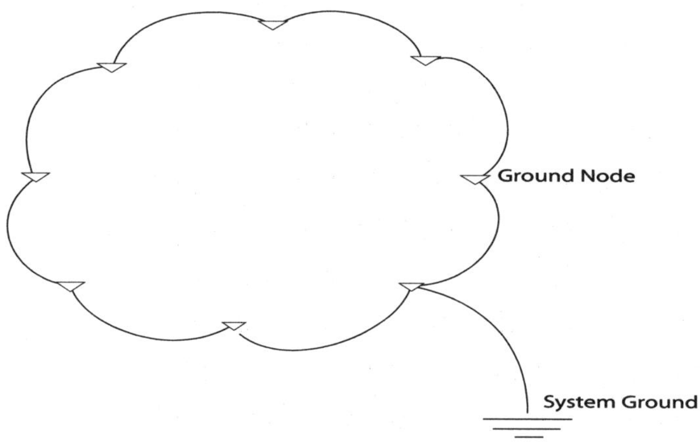

Figure 2.8 Ground Loops

It is also necessary to ensure that there is adequate electrical service to source the interrogator and any auxiliary equipment. Although an interrogator may only require one amp of current, I/O devices can require more, and it is not unlikely that a few hundred watts of power will be needed. When a system suffers brownouts, the interrogator may reset, and there is potential for tag information to be lost.

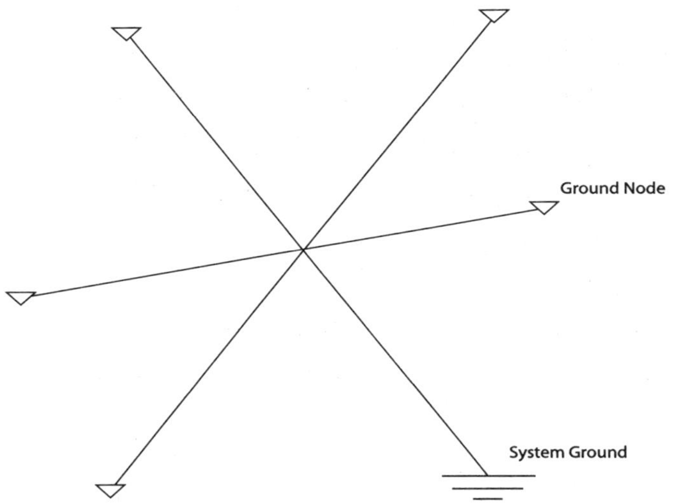

Figure 2.9 Common Ground

Human Safety

RF signals contain energy—the same type of energy used in microwave ovens—so there is the potential for injury *if* power levels are high enough. Regulations establish maximum levels of power generation that have been deemed safe, but observing the variation of power levels specified by different regulatory agencies reveals the lack of consensus on just what is safe. General safety precautions for RFID are similar in any RF system:

- Do not exceed the recommended power levels established by the manufacturer. Regulations do not require the user of RFID equipment to be licensed, but the manufacturer has to establish a safe operating design in order to gain permission to market a product for use in a specific country. Do not "adapt" an interrogator in order to increase power out.

- Do not use any antenna that has not been recommended by the manufacturer. A highly directional antenna may concentrate RF energy in such a way as to exceed safe limits.

- Do not intentionally permit long exposure to the RF energy close to an antenna. RF power is concentrated near the antenna, and this higher power density has more potential for harmful effects on humans than the lower power density away from the antenna.

- Local regulations or laws may require signs that inform people that RF equipment is in use. Check local ordinances.

- Be aware of installations with existing mission-critical RF systems. Existing RF systems will appear in the site survey. If an existing wireless network is carrying critical data (as might be the case in a hospital or clinic environment), extra care must be taken to ensure a new RFID system does not interfere.

- Some new **medical implants** may be programmed using short-range RF signals, and if a patient comes in close contact with an RF signal, the implant may malfunction. This does not mean that all patients are at risk from all RFID installations—only that patients with electronic implants should not be exposed to high RF energy, which might be present close to antennas that power large interrogation zones.

Equipment Safety

Although interrogators, antennas, and cabling are sensitive, they must often be deployed in harsh environments. Interrogators and antennas must be securely mounted, but in such a way that their operation—especially RF propagation to and from tags—is not disrupted. It might be necessary to add plastic guards around the RFID components to prevent people or other equipment from jarring or hitting them (either by accident or on purpose). Cabling should be secured inside conduit for protective and aesthetic purposes.

Summary

This chapter has surveyed the lifecycle of an RFID system, including site selection and design, component installation, system operations, and safety issues. A technologist does not design a system's architecture, but he or she often works closely with the architect's instructions and blueprints. The installation of RFID components is a straightforward process, but numerous technical, security, and safety concerns must be considered before components are mounted and operated. All subsequent chapters of this book explain RFID operations, components, and standards within the framework of the system lifecycle.

Review

Questions

Answers are provided in Appendix A.

1. Which of the following will a blueline drawing not indicate?
   a. Antenna radiation patterns
   b. Doors and windows
   c. Existing RF system installations
   d. Power and network outlets

2. A site analysis will determine _____.
   a. the best placement of RFID interrogator antennas
   b. potential sources of RF interference
   c. readability of a specific tag in the system where it will be used
   d. where to place tags on an item to be tracked

3. A spectrum analyzer shows _____.
   a. voltage in dB versus time
   b. power in watts versus frequency
   c. voltage or power in dB versus frequency
   d. power in watts versus time

4. A "Full Faraday Cycle" analysis is to be performed to prepare for an RFID installation that will use tags that operate at 13.56 MHz. The best frequency span setting for the spectrum analyzer is _____.
   a. 10 MHz, centered at 10 MHz
   b. 3 MHz, centered at 13.5 MHz

c. 13.56 MHz, centered at 6.75 MHz
d. 14 MHz, centered at 13.56 MHz

5. Cables are installed inside conduit because ______. (select all that apply)
   a. more cables can be held in a smaller space using conduit than other methods
   b. conduit shields the cables from RF interference
   c. cables can pose a tripping hazard
   d. the National Electric Code requires it

6. In RFID systems using passive tags, the operation starts with a "wakeup" signal generated by the ______.
   a. tag antenna
   b. interrogator antenna
   c. tag
   d. interrogator

7. Select all the potential sources of ESD from the list below:
   a. Antennas
   b. Boxes moving down conveyors
   c. Grounding cables
   d. Label dispensing
   e. Laper handling
   f. Wireless LAN hubs

8. What component is most sensitive to ESD?
   a. Tag
   b. Interrogator antenna

c. Interrogator

d. Network hub

9. List at least three sources of possible RF interference that would be identified on a blueline drawing as preparation for an RFID installation design.

10. True or False: One shift of operation will be adequate time for a Full Faraday Cycle analysis.

11. Where should antenna be mounted to collect data for a site analysis? (select all that apply)

   a. In the middle of the proposed interrogation zone
   b. On all four corners of the proposed interrogation zone
   c. Random places around the site
   d. Corners of the facility

12. Describe how an antenna and a spectrum analyzer may be used to locate a source of RF interference.

13. Which of the following would be the preferred method of mounting an antenna so as not to interfere with its radiation pattern?

   a. Plastic tie-wraps
   b. Plastic screws
   c. Metal bolts
   d. Metal brackets

14. What current draw would be expected for an RFID interrogator and associated hardware?
    a. 100 mA
    b. 1 A
    c. 3 A
    d. 100 A

Discussion

1. Describe what effect a ground loop can have on system operation.

2. Describe the differences between ground loop and common ground configurations.

3. True or false? All medical devices are immune to RFID systems when installed correctly.

4. Explain how the Full Faraday's Cycle Analysis is used to assess an RFID site. Include an explanation of how this process troubleshoots the interrogation zone. How long should this analysis take?

5. The VSWR ratio was explained in the Primer. What can the VSWR indicate about a system's interrogation zone?

Chapter 3
Frequency Ranges

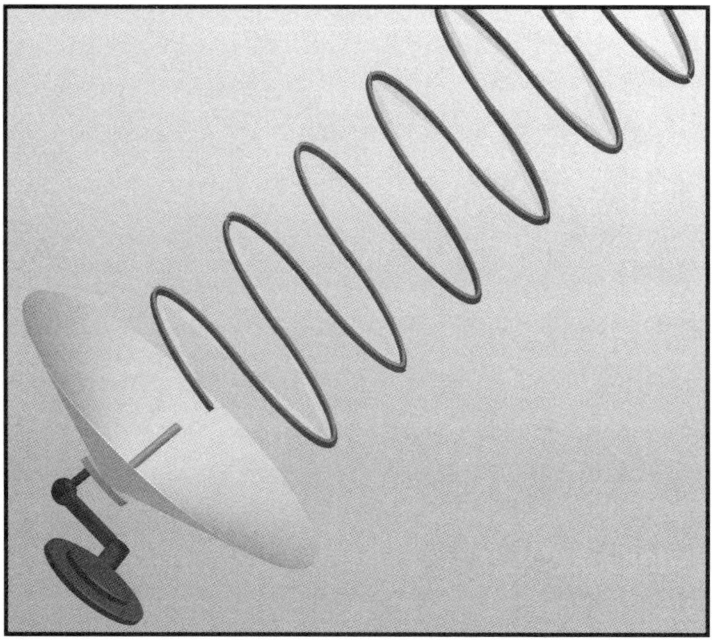

Chapter 3: Frequency Ranges

Key Concepts

- RFID operating frequencies
- Low, high, ultrahigh, and microwave frequencies
- The relationship between frequency and system performance

Terms

Low frequency

High frequency

Ultrahigh frequency

Microwave frequency

RFID systems operate in four main ranges of the frequency spectrum: low, high, ultrahigh, and microwave (Figure 3.1). Frequency does not affect how components operate, but it does affect a system's speed, range, and accuracy.

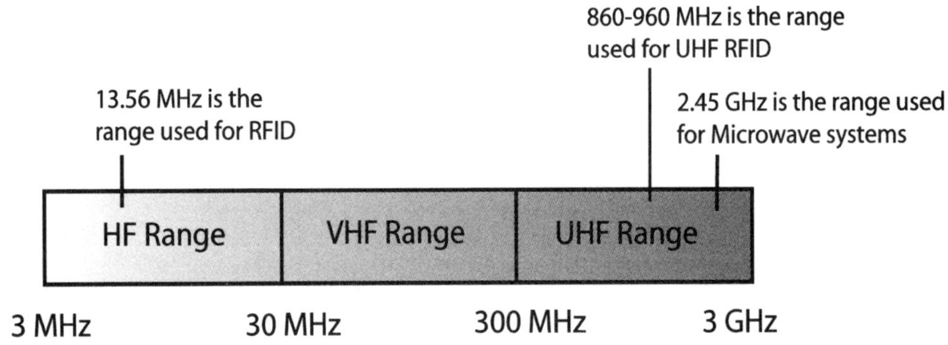

Figure 3.1 Frequency Ranges

Any RFID system in use today has unique advantages and disadvantages. Due to the physics of RF and structure of RFID systems, one part of a system that may be beneficial in certain circumstances is actually disadvantageous in other circumstances. The same situation exists for the frequency ranges within which systems operate. For this reason, it is important that technologists appreciate the trade-offs for each of the four ranges (Table 3.1). Fortunately, system specifications allow a technologist to choose a frequency range that best fits a system application's needs. Anyone with a basic knowledge of RFID technology can haphazardly design a system, but technologists understand what works best in precise situations. This distinction in design and implementation is often the difference between a satisfactory system and a superb system.

Table 3.1 RFID Frequency Ranges

Frequency Range	Band	Read Range	Advantages	Disadvantages	Applications
Low frequency	125 KHz-134 KHz	Below .5 meter	Operates well around water and metals; accepted worldwide	Short read range; slow read rates	Animal tracking Access control Vehicle immobilizers Product authorization
High frequency	13.56 MHz	Below 1 meter	Low cost of tags; accuracy; quick read rates	Require a higher power	Item tracking Airline baggage Smart Cards Libraries
Ultrahigh frequency	860 MHz-930 MHz	3 meters	Frequency selected by EPCglobal	Does not operate well near water or metals	Supply chain and logistics Automated toll collections Parking lot access
Microwave frequency	2.4 GHz	1 meter	Fastest read range	Does not operate well near water or metals	Supply chain and logistics Airline baggage Electric toll collections

Low Frequency Range

Low frequency (LF) range RFID systems are generally used for close contact operations. The LF range extends from 30 to 300 kHz (Figure 3.2). RFID systems within this range operate at 125 KHz and 134 KHz. The read range of these systems is very short (around 20 inches or 50.8 centimeters). Often LF transponders read just beyond the actual contact zone of the interrogator, leaving little room for interception, which increases the system's accuracy and security.

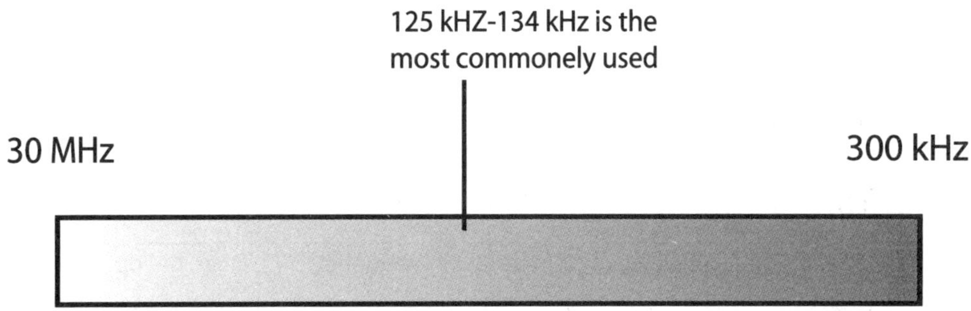

Figure 3.2 Frequencies Used on the LF Range

LF signals are not easily absorbed, so they bounce around in the atmosphere and move through material until an antenna receives them. For this reason LF systems are ideal for use with tools, machinery, and other materials that cause problems for higher frequencies. However, while an LF system can penetrate materials better than high frequency (HF) and other systems, it is inconvenient because it runs with a lower clock cycle, which slows its operation. LF systems also have problems with intensity compromise. Because the range between the interrogator and transponder is small, a higher power intensity can be transmitted and received, leaving more power available for the transponder. The global standards and regulations used for LF systems only permit close range operation.

> **Exam Note**
>
> Do not forget that the antenna power of any RFID system regulates range.
>
> Do not confuse LF range RF principles with RF in a LF RFID system.

The most common applications of LF systems include access control, personnel tracking, animal tracking, automation, and vehicle immobilization security (Figure 3.3). Animal tracking is the most widely used LF application; however, as RFID technologies expand and evolve more LF systems will be implemented for various uses. This makes it is difficult to predict which applications will be used most frequently.

Figure 3.3 Low Frequency Range Application

High Frequency Range

High frequency (HF) RFID systems are used for mid-range distances. The HF range on the frequency spectrum chart is 3-30 MHz (Figure 3.4). The read range of HF RFID systems is approximately 3 feet (1 meter). The actual frequency used is 13.56 MHz (commonly referred to as "thirteen-five-six" by technicians). HF systems have been more broadly adopted because HF is a standard international frequency. It has the capability of operating at a high or medium speed and can be used for low- or high-end RFID systems. The higher frequency enables microprocessors to be used effectively. This frequency provides an RFID user with greater options and speed.

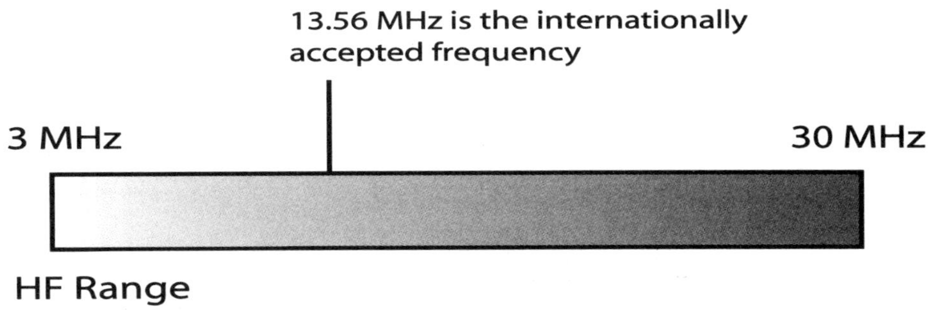

Figure 3.4 Frequencies Used on the HF Range

The trade-offs for HF systems involve two main frequency properties. A higher frequency cannot penetrate objects or materials as well as low frequencies. The other issue in HF systems is tag placement, which can diminish communication between the interrogator and transponder. Tag placement is an increasingly important issue as high frequencies are used in RFID systems. The 13.56-MHz RFID system has the advantage of speed and accuracy (if the tag is placed correctly), but it cannot cover a large range.

Similar to LF systems, HF applications can be used for access control and personnel tracking. However, unlike LF systems, HF systems can handle more tasks including Smart Cards, item tagging, libraries, and EAS (electronic article surveillance) (Figure 3.5). EAS is the RFID system that is often seen at the entrances and exits of retail stores. Additionally, there are many other HF applications in use today or possible in the future.

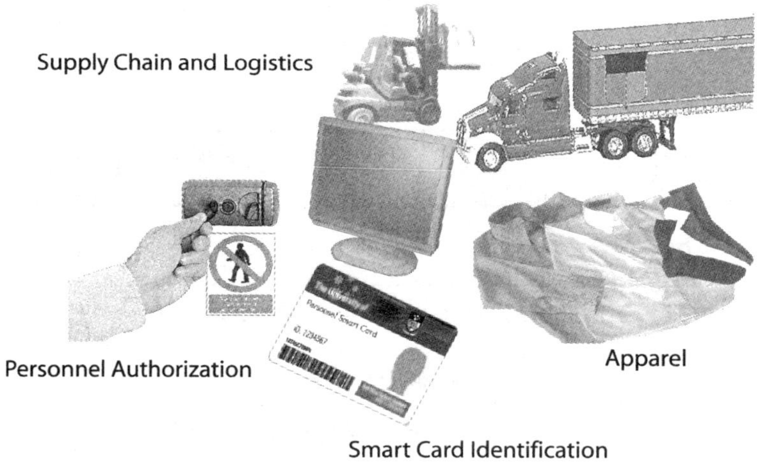

Figure 3.5 High Frequency Range Application

Ultrahigh Frequency Range

The **ultrahigh frequency (UHF)** range covers between 300 MHz and 4 GHz on the radio spectrum. These frequencies are ideal for long-range RFID systems because they have a range of approximately 20 feet (6 meters), and they allow data transfer at even higher speeds. The actual frequencies being used for RFID systems in this range are 433 MHz and 860 to 960 MHz (Figure 3.6). Such systems are among the newest to RFID technology.

The Wal-Mart, the U.S. Department of Defense, and other RFID mandates give users of this frequency band the best opportunity to reduce tag cost. In order for UHF use to become widespread in the future, however, a few issues must be addressed. Primarily, UHF systems experience compatibility problems because regulations for UHF vary by region. For example, a UHF RFID system in the United States will likely not work in Europe. Additionally, the high speed of UHF systems is often detrimental to accuracy. When a system is reading at high speeds, there are more chances for error. UHF systems must be consistently tuned and maintained to avoid technical problems. UHF RFID systems also have problems with interference with many consumer devices that operate in the same frequency spectrum. Interface compatibility is particularly a problem because UHF systems are integrated with a large number of other electronic devices.

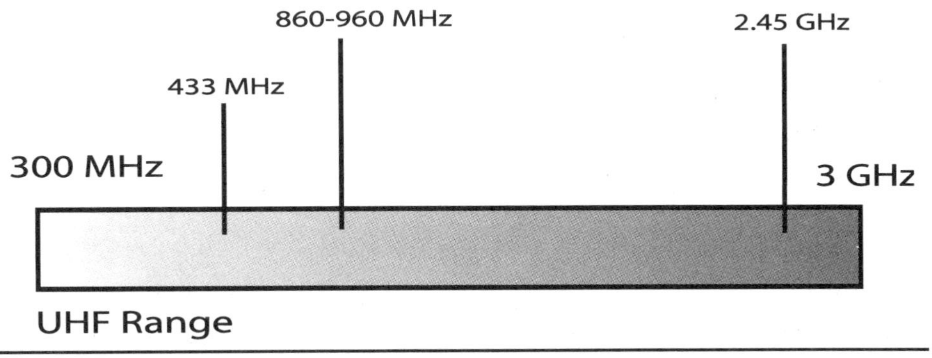

Figure 3.6 Frequencies Used on the HF Range

UHF RFID applications primarily fall into the business arena: logistics, supply chain, and toll collection (Figure 3.7). Warehouse management and inventory tracking are the major areas in which UHF systems are deployed. A company must know what it has, what it needs, and who is buying to ensure productivity. RFID systems in this frequency range allow businesses to access this information quickly and efficiently. For example, UHF toll collection applications make things simple and easy for the people on both ends of the process. A tag is attached to a vehicle and the vehicle is driven through the toll station. No manual actions are required, yet a precise record of time and location is recorded by the application. A destination is reached on time, the account is debited, and the tollbooth conducts business without having to pay a worker. In this sense, the UHF system in use does not simply comply with mandates, it also reduces labor, gains information, and saves time and money.

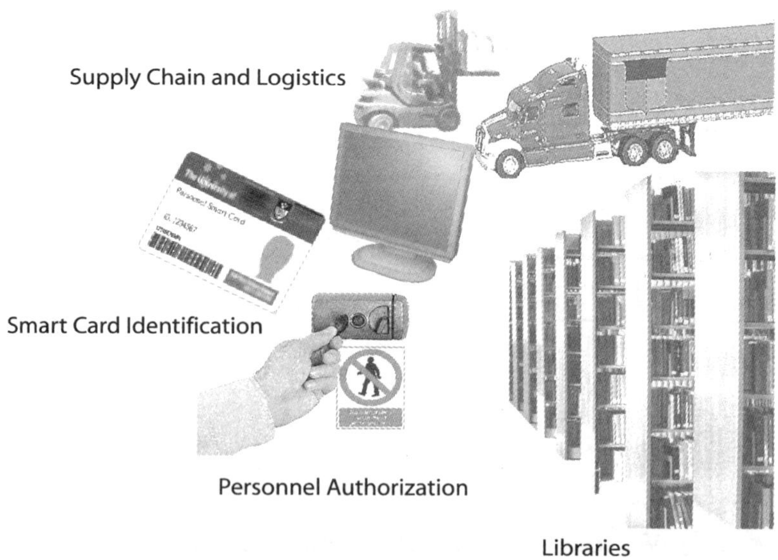

Figure 3.7 High Frequency Range Application

Microwave Range

Microwave frequencies technically fall within the range of the UHF band. Microwave frequencies are located in the upper region of the 300 MHz to 3 GHz frequency band. Typically, these RFID systems are used for long-range verification. Long-range in this case is anywhere between 3 to 10 feet (1 to 3 meters) or longer depending on the environmental setting.

By comparing the UHF specifications and microwave specifications, one can see that higher frequency does not necessarily equal greater range; range is dependant on the power permitted by regulation. The actual frequency used for microwave RFID systems is 2.45 GHz, which has a high data transfer rate.

Because higher frequencies have shorter wavelengths, the transition between near and far field happens closer to the antenna. In the near field, field strength falls off at 60 dB/decade of distance. In the far field, the field strength falls off at only 20 dB/decade. Regulations always give RF power limits at a certain distance from the antenna (different standards might state different distances). If the standard's stated distance for taking power measurements is in the near field, the field strength is going to be much lower just a few feet away. If the measurement is taken in the far field, the field strength will be more consistent on either side of the measurement location. This means that the interrogation zone can extend farther. Additionally, the FCC allows for greater power emissions at higher frequencies. For example, a 915-MHz signal will be limited to 200 microvolts/meter measured at 3 meters, while a 2.4-GHz signal will be allowed to radiate 500 microvolts/meter at the same distance (FCC Part 15 Section 15.209).

Microwaves increase system speed and range. As with UHF, if components are incorrectly set up, microwave systems are inaccurate and intermittent. Like UHF systems, microwave systems have poor performance around water and metals. Microwave systems require line-of-sight or similar propagation to operate. Any propagation obstacles reflect and thus dampen microwave signals. If it deflects off an object, a 2.45-GHz signal has a high possibility of missing the transponder.

Summary

The specific frequencies at which RFID systems operate affect the speed, range, and accuracy of the system's operation. For this reason, the low, high, ultrahigh, and microwave frequency ranges used for operation have advantages and disadvantages according to the physical characteristics of the RF properties in particular ranges. Antennas and other system components must be selected and configured according to a system's operating frequency.

Review

Questions

Answers are provided in Appendix A.

1. RFID "chips" implanted in animals and pets operate in which frequency range?
   a. Low
   b. High
   c. Ultrahigh
   d. Microwave

2. Tags used on the ears of cattle to interface with interrogators about 1 meter away as the cattle walks through portals operate in which frequency range?
   a. Low frequency
   b. High frequency
   c. Ultrahigh frequency
   d. Microwave frequency

3. Low frequency systems are limited to a half meter because _____.
   a. of the RF physics of low frequencies
   b. of the regulations that limit power output of low frequency systems
   c. if frequencies exceed ½ meter they are high frequencies
   d. RFID components cannot transmit at any other frequency

4. Which frequency range operates the best around metals?
   a. Low
   b. High
   c. Ultrahigh
   d. Microwave

5. Which frequency range has the fastest data transfer rate?
   a. Low
   b. High
   c. Ultrahigh
   d. Microwave

6. Which RFID system frequency is internationally accepted?
   a. 134.2 kHz
   b. 13.56 MHz
   c. 915 MHz
   d. 2.5 GHz

7. A library wishes to implement an RFID system. For security and privacy issues, it is desirable that the read range of the tags be approximately 1 meter. The cost per tag should be minimal. Which frequency range should be used?

   a. Low
   b. High
   c. Ultrahigh
   d. Microwave

8. The frequency range that has a read range of 3 meters is the ______.

   a. low frequency range
   b. high frequency range
   c. ultrahigh frequency range
   d. microwave frequency range

9. The 125–134 kHz frequency region is allowed in which of the following areas?

   a. The United States
   b. Canada
   c. Japan
   d. Europe

10. The ______ frequency range is accepted worldwide and operates well around water and metals.
    a. Low
    b. High
    c. Ultrahigh
    d. Microwave

11. Systems on the ________ frequency range are generally used for toll collection and logistics applications.
    a. Low
    b. High
    c. Ultrahigh
    d. Microwave

Discussion

1. Do frequency ranges affect how RFID components operate? If not, why not? In addition, how else might frequency ranges affect RFID systems?

2. What are the safety considerations for systems operating at each frequency range?

3. Identify the advantages and disadvantages of each range of the frequency spectrum. What are some specific system applications that work well at specific frequencies?

CHAPTER 4
ANTENNAS

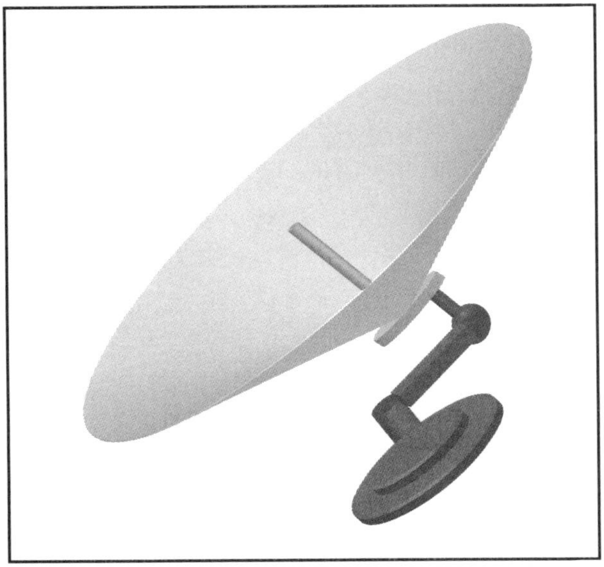

Chapter 4: Antennas

Key Concepts

- Basic antenna function
- Transmission and reception
- Types of antennas
- Radiation patterns
- Adjustment techniques
- Antenna efficiency

Terms

Antenna gain
Backfire antenna
Bandwidth
Beamwidth
Blade antenna
Directional antennas
Directive gain
Directivity
Dome antenna
Effective Isotropic Radiated Power (EIRP)
Far field
Field intensity
Input impedance
Integrated antenna
Near field
Omnidirectional antennas
Panel antenna
Parabolic antenna
Polarization
Power gain
Radiation efficiency
Radiation resistance
Return loss
Sector antenna
Voltage Standing Wave Ratio (VSWR)
Yagi-Uda antenna

Antennas in RFID systems play a vital role in the communication between interrogators and tags. Understanding how different antennas perform is important for selecting or designing antennas for RFID systems. For the purposes of CompTIA's RFID+ exam, a practical understanding of antenna operation is necessary to assess and configure RFID components with consideration for the interrogation zone and the environment.

Basic Operation

An antenna can be defined as the region of transition between a guided wave (transmission line) and a free-space wave (radiation). With transmission line operation, a perfect VSWR of one occurs when the load is equal to the impedance of the circuit, which means all of the incident power is being transferred to the load (Figure 4.1).

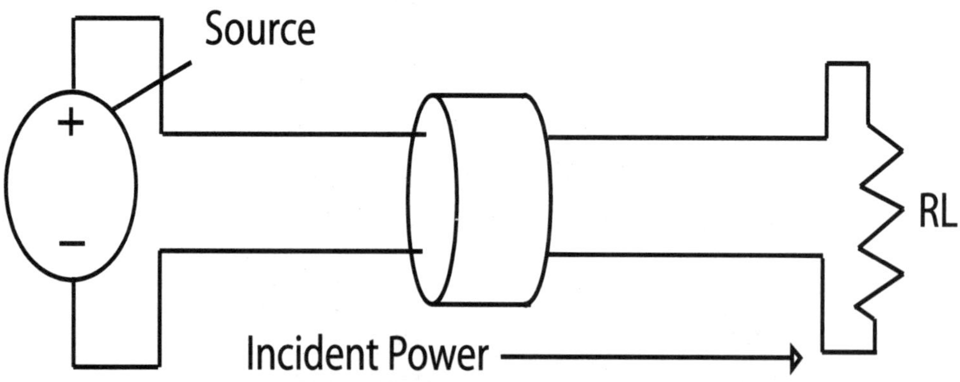

Figure 4.1 VSWR of One

Conversely, if the load is an open circuit, most of the incident power is reflected back toward the source, while a small amount of the power is radiated away from the source, with the open ends of the transmission line acting as radiators or conductors (Figure 4.2). If a person were to bend the wires outward until they were perpendicular to the transmission line, he or she would actually create a simple dipole antenna.

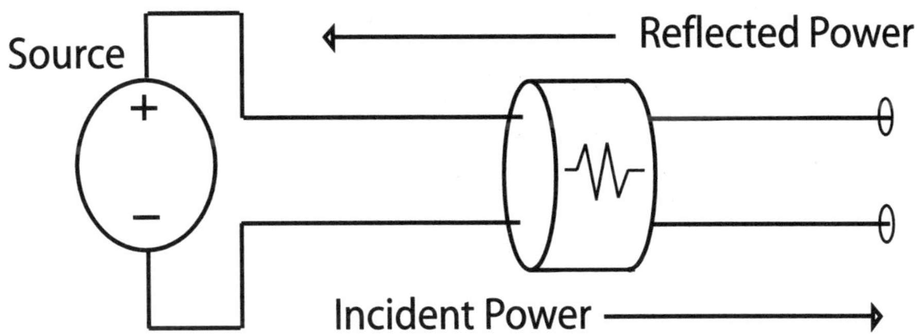

Figure 4.2 Transmission Line Acting as a Radiator or Conductor

Like VSWR, which is a ratio of the power transmitted and received by a transmission line, the amount of energy an antenna radiates and receives can be measured. **Radiation efficiency** is the ratio of radiated energy to reflected energy. In an open circuit transmission line, since the amount of radiated energy is small compared to the amount of reflected energy, the radiation efficiency is low. Spreading the conductors farther apart (like the dipole antenna in Figure 4.4) allows the circuit to radiate more energy. Because

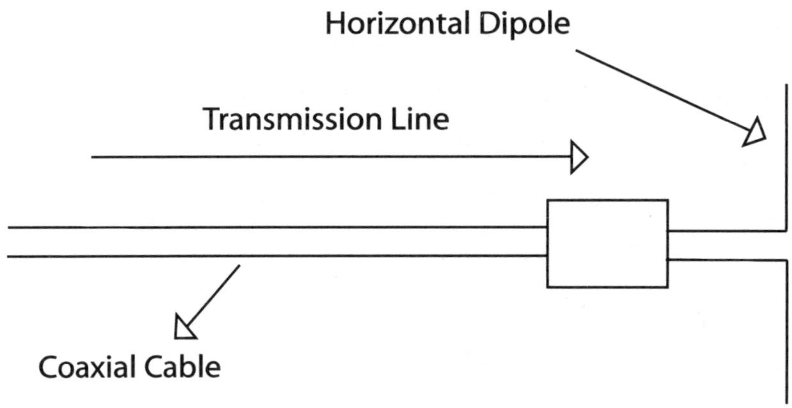

Figure 4.4 Dipole Antenna

antennas are symmetrical in their ability to transmit and also receive signals using the same components, they are considered to be reciprocal devices. Antennas are passive reciprocal devices because they do not actually do any active amplification.

Antenna Radiation Patterns

Each antenna's unique properties govern how it will radiate. Radiation patterns are polar diagrams that represent the variation of field intensity of an antenna with respect to the axis. These patterns, or plots, give a graphical representation of how an antenna responds. Antennas are classified by their radiation patterns into one of two types: directional or omnidirectional.

Directional antennas radiate in a specific direction, while **omnidirectional antennas** radiate uniformly in all directions. Directional antenna patterns are recognized by their front (major) and side (minor) lobes.

The major lobe is also called the front lobe because it propagates (and receives) the most energy at the front of the antenna. Minor lobes represent radiation in a direction different from the major lobe and that reduce the main lobe's intensity. Minor lobes are also called side lobes and back lobes depending on their orientation to the front lobe (Figure 4.5a,b). Radiation patterns for omnidirectional antennas are represented as circles, or more accurately, spheres. When viewing these radiation patterns, it is important to notice that the radiation from an antenna is three dimensional, despite the two-dimensional plots used to depict it.

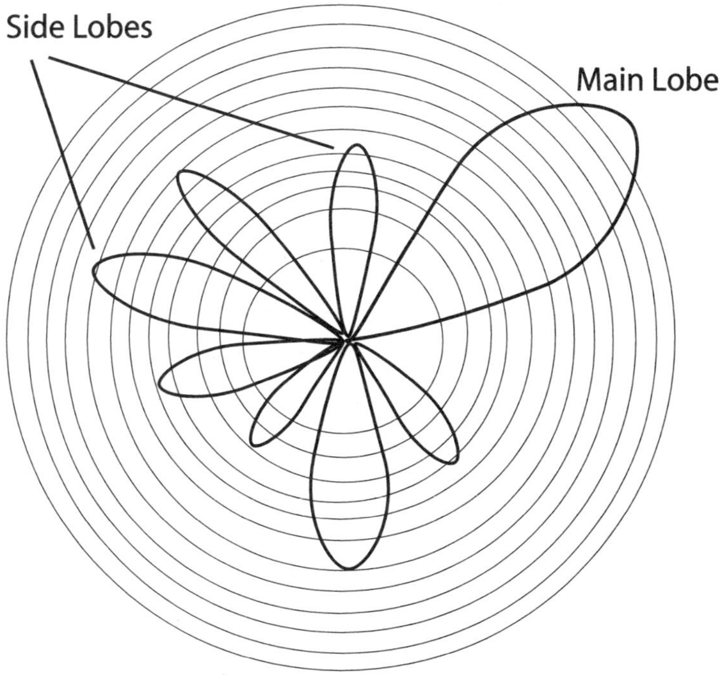

Figure 4.5a Two-Dimensional Radiation Patterns for Directional Antennas

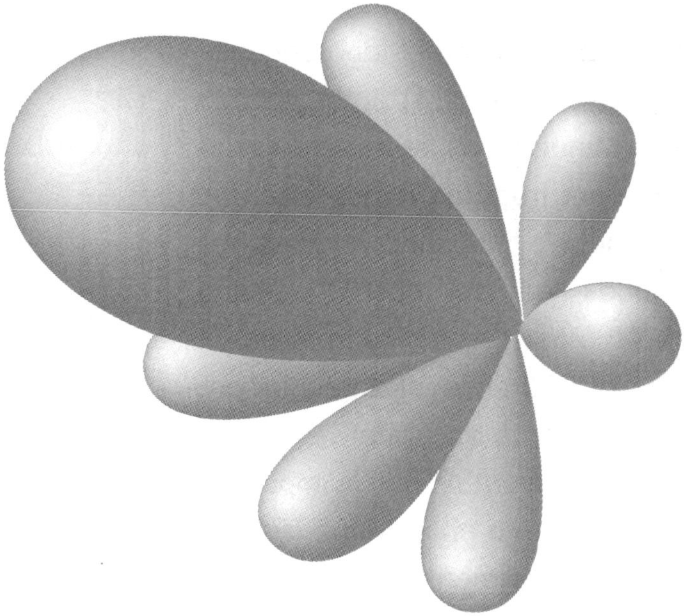

Figure 4.5b Three-Dimensional Radiation Patterns for Directional Antennas

Radiation patterns can also be expressed in terms of **field intensity** or field strength. Field intensity is the vector sum of all the force an antenna exerts at a given point in the transmission field. The pattern that is close to the antenna is called the **near field**, while the pattern that is sufficiently far from the antenna is called the **far field**. When power is radiated from an antenna, some of the power is temporarily stored in the near field and is eventually returned to the antenna. If the receiving antenna is inside the near field, inductive coupling will occur. Power that is radiated into the far field continues to radiate outward and does not return to the antenna. Radiative coupling or backscater is how energy is passed to a receiving antenna in the far field.

As all electrical circuits do, antennas encounter a form of opposition, which is called radiation resistance. **Radiation resistance** is the resistance that would consume the same amount of power that is radiated by the antenna. Radiation resistance cannot be measured directly, but can be calculated by the ratio of power radiated to the square of antenna current at the feedpoint.

$$(4-1) \qquad R_r = \frac{P_{rad}}{i^2}$$

P_{rad} = power radiated (in Watts)
i = antenna current at the feedoint (in Amperes)

All antennas suffer power loss. Antenna efficiency is the capability of radiating effectively with minimum loss. Antenna efficiency is calculated as the ratio of the radiated power to the total input power, where the total input power is the sum of radiated power and the power dissipated.

$$(4-2) \qquad \eta = \frac{P_{rad}}{P_{in}} \times 100\ \%$$

Where $P_{in} = P_{rad} + P_{dis}$ (in Watts)

Antenna Gain

Antenna gain is a measure of the ability of the antenna to focus radio waves in a particular direction. The gain of the antenna depends on the antenna's size, operational frequency, and efficiency. Directive gain and power gain are two terms that are often misunderstood. Assuming both antennas are radiating the same power, directive gain is the ratio of power density radiated in a particular direction by a given antenna to the power density radiated by a (omnidirectional) reference antenna to the same point. This is defined as:

$$(4-3) \qquad D = \frac{P}{P_{ref}}$$

P = power density at a specific point with a given antenna (in Watts)

P_{ref} = power density of a reference antenna at the same point (in Watts)

Maximum directive gain is referred to as **directivity**. Directivity is a measure of how focused an antenna coverage pattern is in a given direction. Because directive gain is based on the shape of the far-field pattern, the antenna efficiency is not taken into account.

Power gain accounts for both the directive property of the antenna and the antenna's efficiency in transforming input power into radiated power. Assuming the given antenna and the reference antenna have the same input power and the reference antenna is lossless, power gain can be described mathematically as:

$$(4-4a) \qquad A_p = D\eta$$

D = directivity

η = efficiency

If the reference antenna is a lossless isotropic, it radiates all of the input power, thus power gain is equal to directive gain. Power gain can be measured in decibels and is defined as:

$$(4-4b) \qquad A_p = 10 \log (D\eta)$$

There is a common misconception that when antennas exhibit gain they automatically create power within themselves. This is not the case. Power radiated from an antenna can never exceed the input power. An antenna does not actually amplify the power, but rather focuses the radio waves into narrower patterns so that there appears to be more power coming from the antenna. Without any increase in power overall, increasing power in one direction must result in decreasing power in other directions.

Effective isotropic radiated power (EIRP) or simply effective radiated power (ERP) at a given point in space is the power that would have to be supplied to an isotropic antenna to achieve the same field strength or power density as another antenna at that point. EIRP can be expressed mathematically as:

$$(4-5a) \qquad EIRP = P_{rad} \bullet D_t$$

P_{rad} = total radiated power (in Watts)

D_t = transmit antenna directive gain

It can also be expressed in decibels relative to one milliwatt as:

$$(4-5b) \qquad EIRP_{(dBm)} = 10 \log \frac{P_{rad}}{0.001\ W} + 10 \log D_t$$

Or in decibels relative to one watt as:

$$(4-5c) \qquad EIRP_{(dBW)} = 10 \log (P_{rad} \bullet D_t)$$

Using antenna input power and power gain in place of radiated power and directive gain, equation 4=-5a can be rewritten as:

$$(4-6a) \qquad EIRP = P_{in} \bullet A_t$$

P_{in} = total antenna imput power (in Watts)

A_t = transmit antenna power gain

It can also be expressed in decibels relative to one watt as:

$$(4-6b) \qquad EIRP_{(dBm)} = 10 \log \left(\frac{P_{in} \bullet A_t}{0.001\ W} \right)$$

Or in decibels relative to one milliwatt as:

$$(4-6c) \quad EIRP_{(dBW)} = 10 \log\left(\frac{P_{in} \bullet A_t}{1\,W}\right)$$

To determine power density at a given point, R in the field, from a transmit antenna, Equation 4-4a can be written to include the transmit antenna gain as:

$$(4-7a) \quad P = \left(\frac{P_{in} \bullet A_t}{4 \bullet \pi \bullet R^2}\right)$$

P = power density

P_{rad} = power radiated from transmit antenna

A_t = transmit antenna power gain

D_t = transmit antenna directive power gain

R = distance from the transmit antenna

Or in terms of directive gain:

$$(4-7b) \quad P = \frac{P_{rad} \bullet D_t}{4 \bullet \pi \bullet R^2}$$

Captured Power Density

Since antennas are reciprocal devices, they have the same characteristics during transmission as they do during reception. These characteristics include radiation resistance, efficiency, power gain, and directivity. This reciprocity enables the calculation of captured power density, C, which is the power density in space, with the following equation:

$$(4-8) \qquad C = \frac{P_{in} \cdot A_t \cdot A_r}{4 \cdot \pi \cdot R^2}$$

C = captured power density

P_{in} = transmit antenna input power

A_t = transmit antenna power gain

A_r = receive antenna power gain

R = distance between the transmit and receive antenna

Power

More important than captured power density is the actual power that a receiving antenna produces at its output terminals. This power is dependent on the amount of power captured by the receiving antenna and the efficiency of the transmitting antenna.

$$(4-9) \qquad P_{cap} = P \bullet A_c$$

P_{cap} = captured power

P = power density

A_c = capture area

A receiving antenna captures a portion of the EIRP, which is determined by its effective capture area. The received power or captured power available at the antenna terminals is the power density multiplied by the effective capture area of the receiving antenna.

Capture Area

For a given receiver antenna size, the capture area is constant no matter how far it is from the transmitter.

There is a relationship between an antenna's size and its ability to capture electromagnetic energy as well as a relationship between the antenna gain and the antenna's receiving area. This relationship is stated as:

$$(4-10) \qquad A_c = \frac{A_r \bullet \lambda^2}{4 \bullet \pi}$$

A_c = effective capture area
P = wavelength of receive signal
A_r = power gain of receive antenna

Polarization

Polarization is the direction in which the antenna radiates wavelengths; it can be vertical, horizontal, circular, or elliptical. When the RF signal leaves a dipole antenna, it is aligned, or has linear polarization. The wave front from the antenna radiates outward, but always parallel to the antenna. This means that the receiving antenna must be positioned parallel to the transmitting antenna in order for signal to be picked up (Figure 4.6).

Using a dipole antenna with an interrogator will not work if the tags are not aligned with the transmitting antenna. One popular solution is to use two antennas, at a physical 90° angle with each other, thus covering more angles. The configuration is further enhanced if there is a time delay (of a 90° phase angle or quarter wavelength) between the two antennas. The wave front appears to rotate 360° with every cycle, so any tag orientation is matched. Depending on which one of the antennas has the delayed signal, and the orientation of the antennas, the antenna may be classified as right-handed polarization or left-handed polarization. Since the signal is split between both antenna elements, there is a polarization loss of 50% (3dB) with a circular antenna compared linear polarized antenna.

Beamwidth and Bandwidth

Beamwidth measures how energy is focused or concentrated (Figure 4.7). It is simply the angular separation between the two half-power points (−3 dB) on the major lobe of a radiation pattern. **Beamwidth** is inversely proportional to antenna gain: the wider the beamwidth, the lower the antenna gain. Typically, antennas have beamwidths between 30° and 60°. Omnidirectional antennas, however, have a gain of one and a beamwidth of 360° due to their equal distribution in all directions (Figure 4.8).

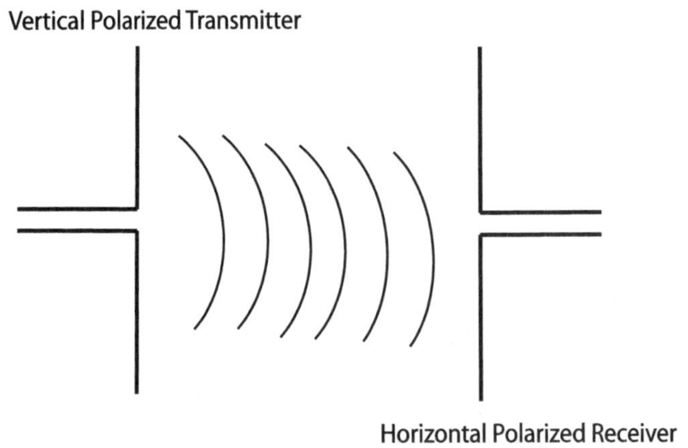

The transmitting and recieving antennas are aligned and the signal is received.

Figure 4.6 Parallel Antennas

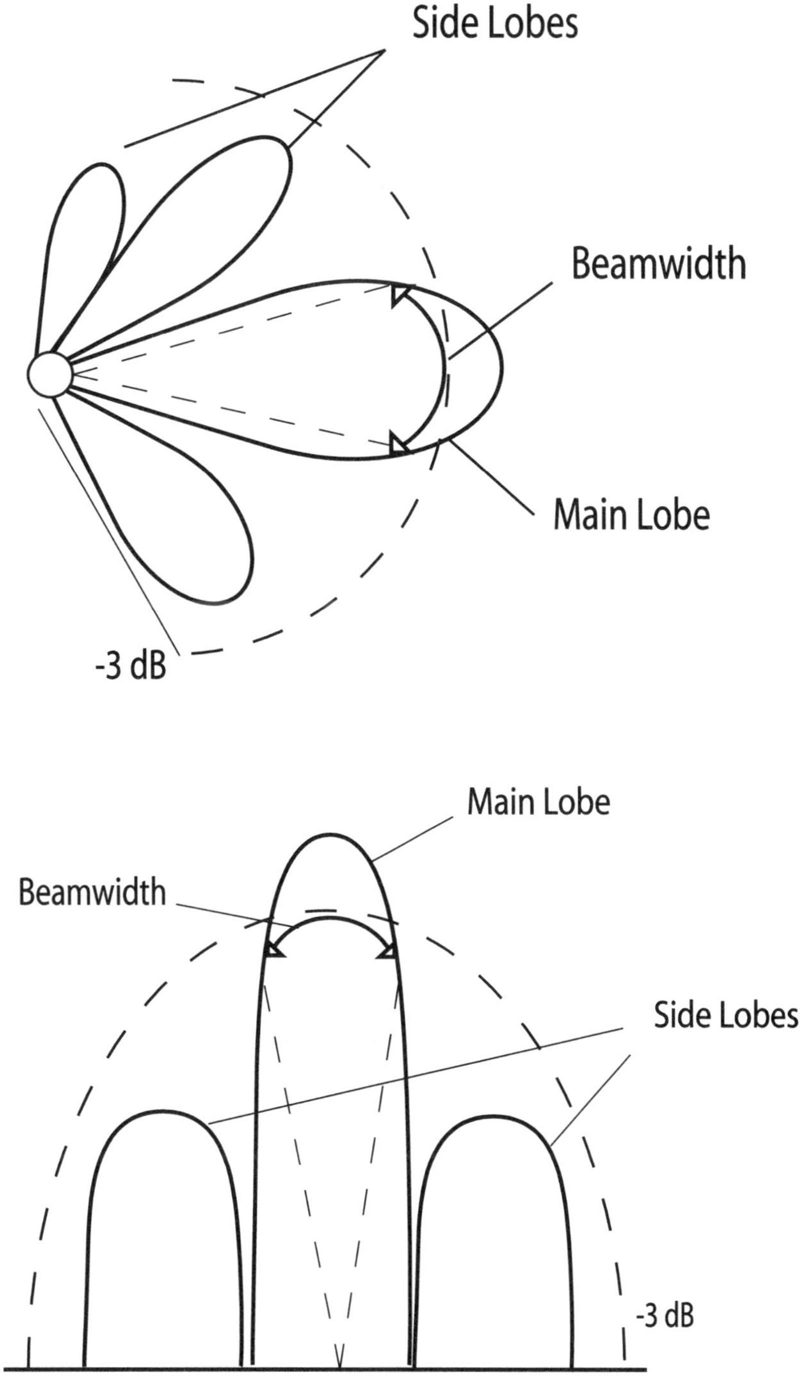

Figure 4.7 Antenna Beamwidth

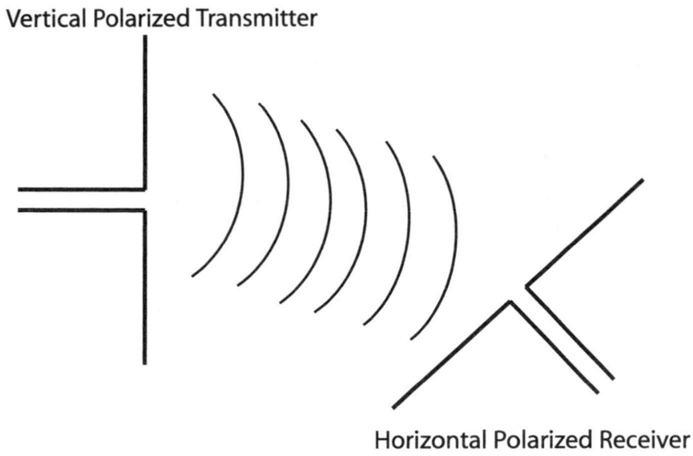

Figure 4.8 Antenna Polarization

Bandwidth primarily refers to the frequency span from the highest to lowest frequency of an assigned frequency band.

$$(4-11) \qquad BW = f_h - f_t$$

f_h = high half-power frequency

f_t = low half-power frequency

A closely related figure of merit is the antenna's Q, which is typically expressed as a percentage of the antenna's optimum frequency of operation and the difference between the highest and lowest frequencies of operation. The higher the Q, the narrower the bandwidth is and the more selective the antenna.

$$(4-12a) \qquad Q = \frac{f}{f_h - f_t}$$

f = optimum frequency of operation

For example, to determine the Q for an antenna with an optimum frequency of operation of 1 MHz and half-power frequencies of 1.2 MHz and 0.8 MHz, you would insert these variables into Equation 4-12b as follows:

$$(4-12b) \qquad Q = \frac{1}{1.2-0.8} = 2.5$$

Input Impedance

Antenna **input impedance** is the ratio of the antenna's input voltage to input current. Current flows to the antenna through a transmission line connected to the antenna at a feedpoint. If the transmitter's output impedance and the antenna's input impedance are equal to the characteristic impedance of the transmission line, maximum power will be transferred to the antenna and radiated since no standing waves will be present on the line. To calculate input impedance:

$$(4-13) \qquad Z_{in} = \frac{E_i}{I_i}$$

Z_{in} = antenna input impedance

E_i = antenna imput voltage

I_i = antenna input current

Voltage standing wave ratio (VSWR) measures how well the antenna is matched to the network at the operating frequency being used. It indicates how much power will be transferred to the antenna and radiated, depending on how many standing waves are present on the line.

Return loss ultimately measures how well matched an antenna is to the network. Return loss is the difference in power between the incident power and the power reflected back by the load due to a mismatch. It is expressed in dB as:

$$(4-14) \quad \text{Return loss} = 10 \log \frac{P_{reflected}}{P_{incident}}$$

A perfect impedance match would result in no reflected power since it would all be delivered to the load; therefore, the return loss would be infinite. Conversely, the return loss on an open circuit would be zero because all of the power would be reflected back. When dealing with return loss, the higher the value, the better the impedance match.

Types of Antennas

Antennas are classified by their radiation patterns into one of two types: directional or omnidirectional. Directional antennas transmit and receive power in a particular direction. Some types of directional antennas are backfire, Yagi-Uda, parabolic, panel, and sector.

Backfire antennas offer excellent gain and are ideal for outdoor use because of their unobtrusive, compact design (Figure 4.9).

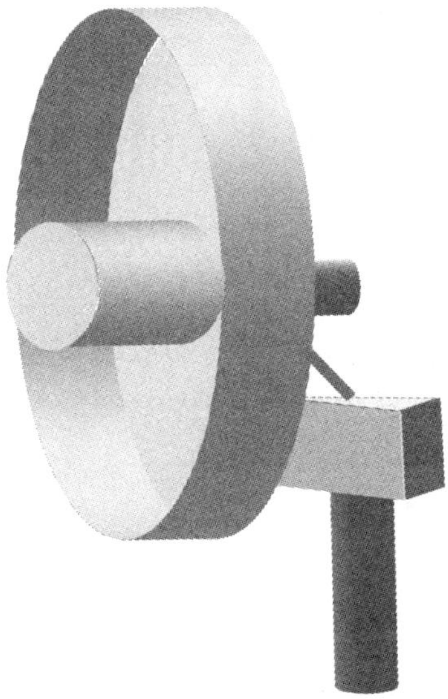

Figure 4.9 Backfire Antenna

The **Yagi-Uda antenna**, named after its Japanese inventors, Professors Hidetsugu Yagi and Shintaro Uda, consists of an array of dipoles of different lengths and spacing in order to maximize radiation in a forward direction (Figure 4.10). Yagi antennas also offer excellent gain and are superb for use in long narrow areas.

Parabolic (dish) antennas use a reflector in the shape of a paraboloid, which has a distinct focal point where a small, low-gain feed antenna is located. The principle of this design is to focus the energy into a narrow beam that can travel long distances (Figure 4.11).

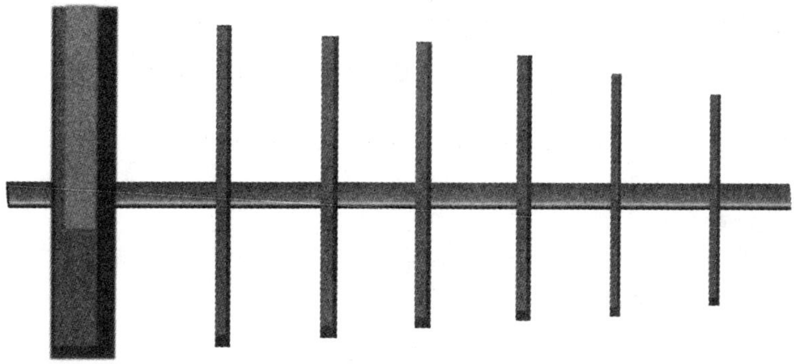

Figure 4.10 Yagi-Uda Antenna

Figure 4.11 Parabolic Antenna

Panel antennas are rectangular-shaped antennas designed to radiate forward and to the sides, horizontally. These are often called "picture frame" antennas because they can be hung on walls and do not radiate backward in the direction of the wall. Panel antennas are ideal for applications such as interrogator zones where direction is important but polarization is not desired (Figure 4.12).

Sector antennas are used in applications that need to direct radio waves over a large, yet specific area. Sector antennas are typically large, dish or flat-panel shaped designs (Figure 4.13).

Figure 4.12 Panel Antenna

Omnidirectional antennas radiate power uniformly in all directions. Some types of omnidirectional antennas are integrated, blade, and dome.

Integrated antennas are built into other devices, commonly wireless networking devices such as wireless network cards or wireless access points. Because integrated antennas are often embedded within other objects they typically cannot transmit and receive as well as external antennas.

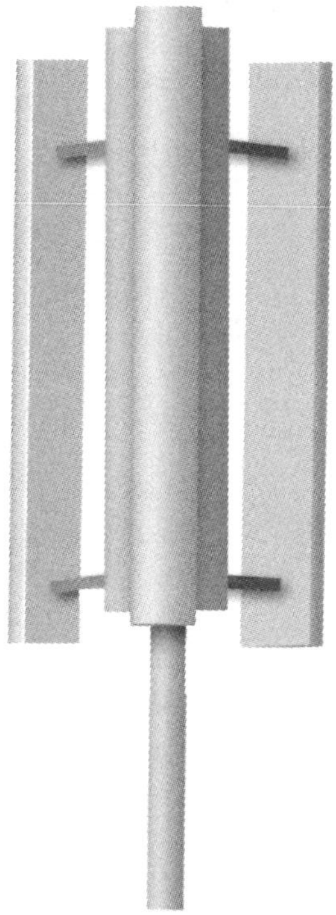

Figure 4.13 Sector Antenna

Blade antennas are shaped somewhat like knifes. These long, thin antennas are usually housed in plastic and are used to boost a signal or to pick up low power signals (Figure 4.14).

Dome antennas, often referred to as ceiling dome antennas, are commonly used in situations that require a wide area of coverage. They look similar to smoke detectors and have the advantage of being unobtrusive since most are mounted on the ceiling. Most corporate offices that utilize cubicle farms have embraced the dome antenna (Figure 4.15).

Figure 4.14 Blade Antenna

Figure 4.15 Dome Antenna

UHF and Microwave Antennas

Microwave antennas have essentially the same operating principles and electrical characteristics as low-frequency antennas. They have highly directive half-power beamwidths of 1° or less. Narrow beam-width minimizes the effects of interference from outside sources and adjacent antennas. A drawback, however, is that microwave antennas require line-of-sight transmission. Additionally, narrow beam-widths have several limitations, such as mechanical stability and fading, which can cause problems in antenna lineup. Similar to directional antennas, microwave antennas do not radiate all electromagnetic energy in the direction of the main lobe. Some energy is directed to the side and back lobes. These side lobes can be prone to interference from or a cause of interference to other microwave signals.

The coupling method for UHF and microwave antennas differs from the inductive coupling discussed earlier, mainly because higher frequencies have shorter wavelengths. In the near field of the antena's signal—within approximately $\lambda/2\pi$ (where λ is the wavelength of the signal) of the antenna—the electromagnetic wave is still interacting with the antenna. For example, a frequency of 27.125 MHz would have a near field extending to 1.7 m. As the wave radiates away from the antenna, it detaches and travels out in a straight path. When two antennas are farther away from each other than $\lambda/2\pi$, inductive coupling will no longer work. In these cases, a mechanism of radiative coupling, or backscatter, is the prevalent mode. The wavefront comes in contact with the receiving antenna after it has detached from the transmitting antenna. In order for backscatter to be efficient, the same requirements for antenna length and impedance matching apply. If the receiving antenna is to resonate with the incoming wavefront, it must be of similar length; therefore, the dipole antennas used with microwave systems, while smaller than their HF counterparts, are still approximately $\lambda/2$ in length.

Summary

Antennas are used to transmit and receive data between tags and interrogators. An antenna is any form of conductor that transmits and receives signals. Understanding radiation, intensity, resistance, and gain enable a technologist to measure and diagnose an antenna's reception and transmission. Specific antenna types modify the direction, beamwidth, and bandwidth of signals, and can therefore be selected for particular settings and purposes.

Review

Questions

1. Which of the following devices can be used to visually display wave signals?

   a. Transmitter

   b. Oscilloscope

   c. Stethoscope

   d. GUI

2. Antenna Gain is a measurement of _____.

   a. the total intensity of an antenna's signal.

   b. a ratio of the power an antenna receives and transmits.

   c. the ability of an antenna to focus radio waves in a particular direction.

   d. the average bandwidth of the waves an antenna propagates.

3. Directivity is also known as?

   a. ERP

   b. field direction

   c. maximum directive gain

   d. directive power

4. The main lobe of directional antenna the _____.
   a. maximum VSWR point of radiation
   b. direction of least radiation from the antenna
   c. direction of maximum radiated field intensity from the antenna

5. An antenna with low radiation resistance will have a high/low efficiency?

6. True or False: Antenna efficiency is always less than 1 (or 100%).

7. True or False: Directive gain is always less than 1.

8. True or False: Power gain is always less than 1.

9. Calculate the power gain in decibels for an antenna with directive gain of 5 and efficiency of .5 (50%).

10. A power specified in dBm means that _____.
    a. the actual power was multiplied by 10-3 before taking the logarithm
    b. the actual power was expressed in mW before taking the logarithm
    c. there is no difference between dB and dBm, so it has no significance
    d. the directive gain in dB was added to the result

11. The EIRP is a measure of _____. (select all that apply)
    a. the directivity of an antenna
    b. the efficiency of an antenna
    c. power density at a particular point in space
    d. power radiated by a specific antenna

12. Calculate the power density 1 m away from an antenna with EIRP of 3 dBm.

13. A receiving antenna has a power gain of 0.5, and cap tures 30 mW of power at 900 MHz. Calculate the actual power density of the 900 MHz signal.

14. How much current is needed to create 100 mV across a 50 Ω antenna? How much power is this?

Discussion

1. An antenna is operating at 10 MHz. If a tag is within 30 cm of the antenna, what type of coupling is taking place? Explain your answer.

2. What type of antenna would be best for the for following applications and why?
    - The tracking of documents at a law firm
    - An EAS system for a power plant
    - Inventory logistics for a small warehouse

3. Describe the differences between inductive coupling and backscatter. Why is backscatter the prevalent mode in microwave systems?

CHAPTER 5
TAGS

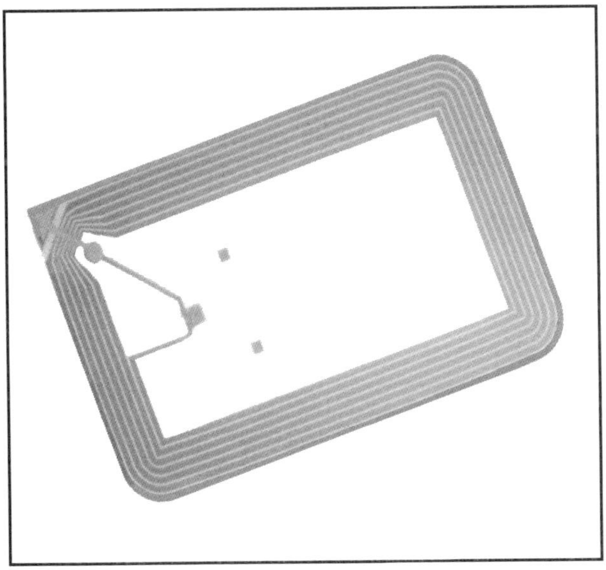

Chapter 5: Tags

Key Concepts

- Tag components and functions
- Tag types, design, and use
- Tag memory
- Tag classes
- Tag placement

Terms

Absorption

Active tags

Disk/Coin tags

Contact-less Smart Card

Integrated circuit (IC)

Passive tags

Read-Only (RO)

Read-Write (RW)

Reflection

Refraction

Semi-passive Tags

Smart label

Standing wave

Write Once Read Many (WORM)

Tags, or transponders, are an integral part of an RFID system. When technologists select and configure tags in a system, they should consider the design and function of tags. Other primary concerns include tag classification and placement. Each of these considerations will reinforce the technologist's capacity for making the best decisions for the overall system.

In system operation, when an interrogator transmits a signal to a tag, the tag's task is to then communicate its identity, as well as the identity of the attached object, back to the interrogator. Once an interrogator identifies an object's tag, it accepts information about the object from the tag.

There are three primary types of tags produced: passive, semi-passive, and active. Passive tags do not have an interior power supply, and therefore they must rely on the interrogator for power. Active tags possess batteries that supply power for communication. For this reason, active tags can significantly improve the range between the tag and the interrogator. A hybrid of the active and passive tag, the semi-passive tag possesses a low-cost battery that is used to power the tag's onboard electronics. The battery does not allocate power for increasing the tag's communication range.

Tag Types

There are numerous vendor-specific tag types, and many of these models are similar in format and design. One common format is the **disk** or **coin tag** (Figure 5.1). With the disk or coin format, a round plastic casing, which usually contains a central hole for fastening, encloses the tag.

Figure 5.1 Disk/Coin Tag

Another format that is increasingly used is the **smart label** (Figure 5.2). A smart label is a paper-thin tag with its antenna coil in the form of foil etched or printed onto the back of the plastic. After it is printed, a smart label is laminated and then coated with an adhesive. This design allows the tag to be affixed to inventory items just as easily as barcode stickers.

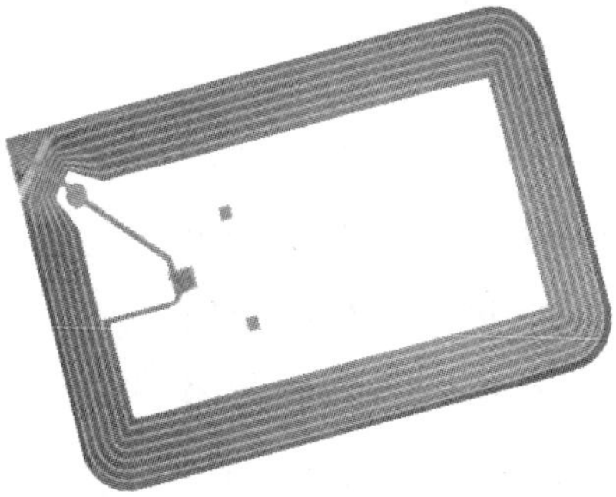

Figure 5.2 Smart Tag

A format already familiar to many people is the **contact-less Smart Card** (Figure 5.3). This format is the same shape and size as a credit card. These tags are created by sandwiching the transponder between permanent vital circuit (PVC) foils. The Smart Card format allows for a large coil area that gives this type of tag an increased communication range. Other available formats include glass-housed, plastic-housed, ferrite pot cored, key ring, watch, and even coil-mounted tags. As RFID matures, research and innovation will likely spur the development of many more formats.

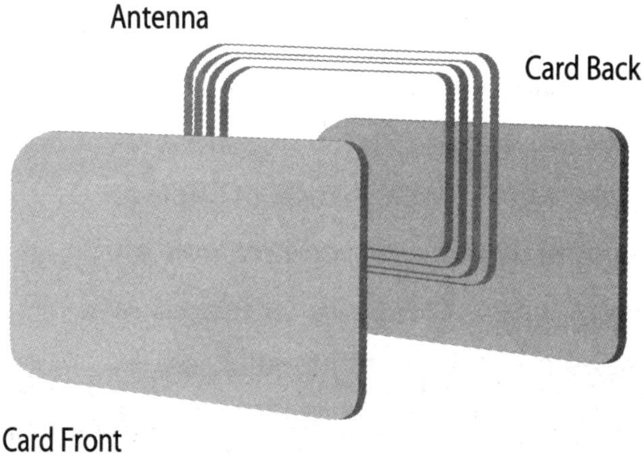

Figure 5.3 Contact-Less Smart Card

Tag Components

A tag consists of an **integrated circuit (IC)** that attached to an antenna (Figures 5.4, 5.5). The IC and antenna are usually placed into some form of protective inlay, which can be as large as a credit card or small enough to fit on a fingertip, depending on the application and tag manufacturer. The alternative to IC tags, called chipless tags, costs less than IC tags but is used mostly for simple tasks such as large-volume inventory monitoring. Because nanotechnology has become an instrumental part of tag design, many tags are astonishingly small, some as small as a grain of sand. The actual size of a tag is determined by the antenna's size. While the IC can be very small, the antenna must be at least long enough to receive and respond to the interrogator's signal. A tag's functionality is determined by its antenna's configuration and length. In receiving and responding to signals, antennas typically need measure at least one-quarter of the original frequency wavelength.

Example

There is a 900-MHz signal. The wavelength of this signal is calculated by using the following wavelength formula:

$$\lambda = \frac{c}{f}$$

$$c = 3 \cdot 10^8 \text{ m/s}$$
$$f = 900 \text{ MHz}$$

The result is a wavelength of approximately 1.09 feet or 33 centimeters. One-quarter of this result yields an antenna length of approximately 3 inches or 8.3 centimeters.

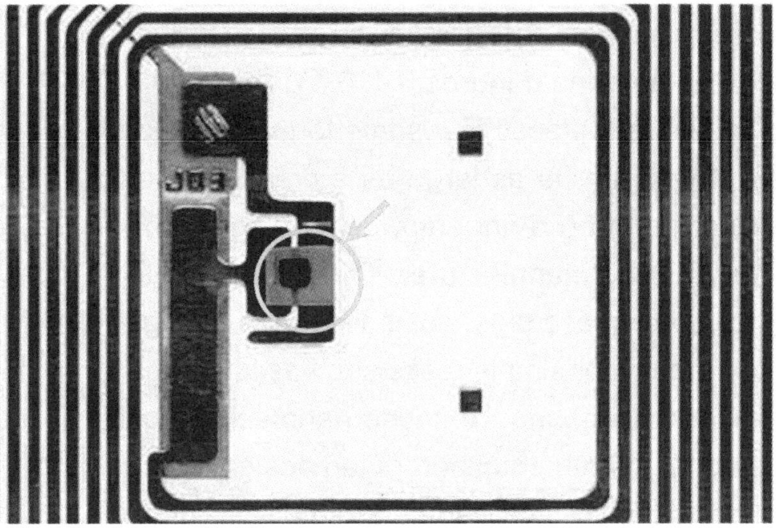

Figure 5.4 Infared Photo of Integrated Circuit

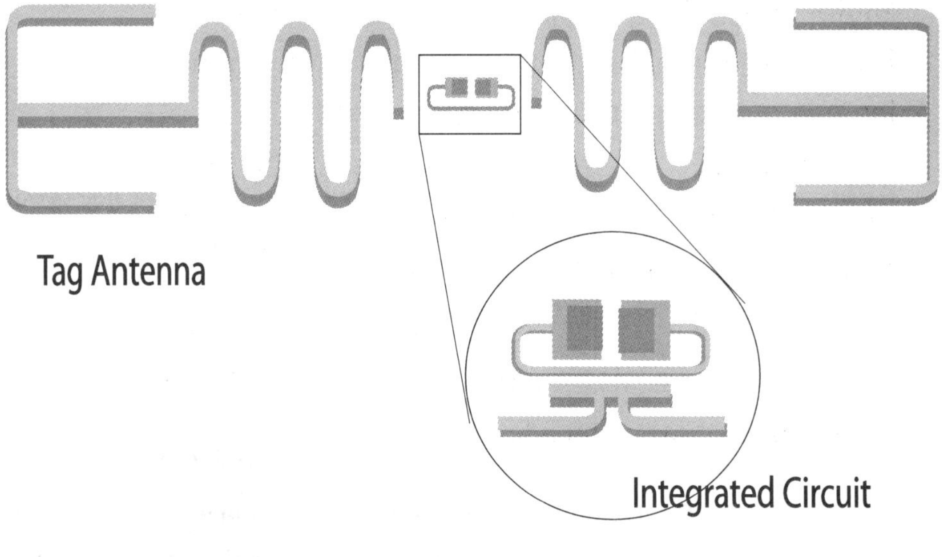

Figure 5.5 Tag and Integrated Circuit

Tag Functionality

Tags have various memory structures and data capacities; therefore, there are numerous methods of retrieving and transmitting information with RFID tags. Tags will transmit and receive information according to the type of information it stores and the specific commands of an interrogator. In systems using first generation protocols, tags from one vendor did not talk to interrogators from another vendor. All types of tag memory can perform basic read-and-write tasks. The RFID tag costs are directly associated with the complexity of memory read-and-write functions.

Tag memory is designed in three ways: **Read-Only (RO)**, **Write Once Read Many (WORM)**, or **Read-Write (RW)**. In terms of complexity and cost, each tag type has advantages and disadvantages according to its memory type. A complex tag (RW) may cost more and provide a greater functionality, but a simple tag (RO) provides more security and ease of use. More information on the read-write functions associated with RFID tags will be discussed later in this chapter. All types of memory can be used in passive, semi-passive, and active tags.

Passive Tags

A **passive tag** is so named because it cannot generate and reflect radio signals to an interrogator if it is not in the presence of an electromagnetic (EM) field. A passive tag must be inside the interrogation zone in order to receive enough power to generate a response. The initial interrogator signal powers a passive tag's circuitry, allowing the tag to function.

Example

The interrogator transmits EM waves. Next, the tag antenna receives the transmitted EM waves, and current is induced. The tag uses the induced current to "talk back" (a.k.a. backscatter) to the interrogator. The backscatter response for a passive RFID tag is an amplitude modulated (AM) signal (Figure 5.6).

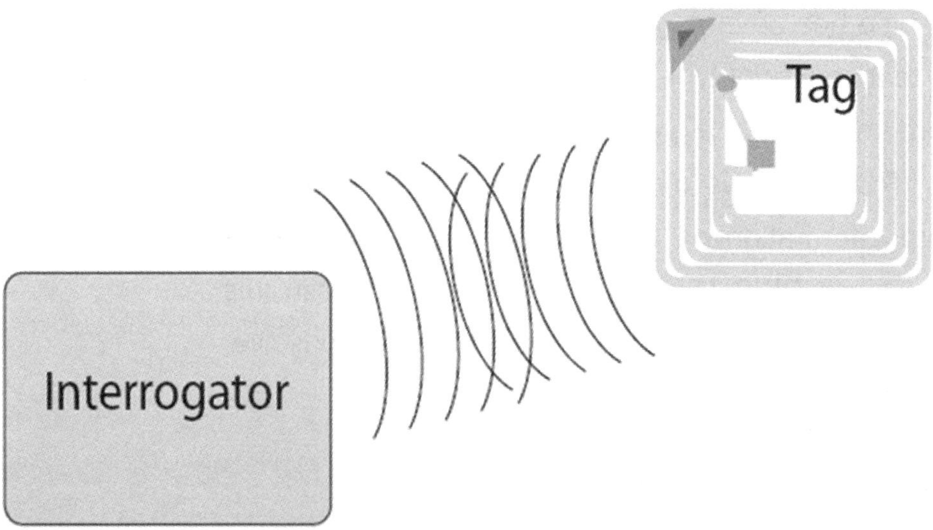

Figure 5.6 Electric Field Backscatter

Passive tags are, by far, the most economical and popular tags in today's RFID industry. For the most part, passive tags are small and have a decreased read range. Furthermore, there is no battery inside the tag, which lowers the cost. Passive tags have a long tag life, since there is no need to replace a battery after a certain period of time. These tags are power efficient, as well, because they are only on while being accessed by the interrogator.

Memory capacity for passive tags varies. For example, EAS (electronic article surveillance) uses 1-bit passive tags. With the passing of more global standards, newer generation passive tags are expected to have memory capacity of 2 kilobytes or more. Beyond the cost, another benefit of passive tags is their numerous design options, a flexibility that makes implementation easier. RFID in the supply chain uses primarily passive tags. With retailers hoping to drive down the cost per tag by using billions of these types of tags, logistics and supply chain applications will utilize passive tags for some time.

Semi-Passive Tags

Semi-passive tags are the least used in the RFID industry, and are truly an "in-between" method of tag operation. As with passive tags, semi-passive tags make use of the initial EM wave produced by the interrogator to generate communication. In this way, the operating principle of the semi-passive and passive tags is the same: the tags only operate when they receive a "wake-up" signal from the interrogator, and can only transmit when they receive adequate power from a source. The tags use their own batteries to run circuitry.

Semi-passive tags' additional battery power allows them to resist interference or circumvent a lack of power from the original interrogator signal. The transmission signal sent back to the interrogator is stronger than that of a passive tag's, which allows tags to transmit across longer distances and sustain operation for the proper amount of time during the information transaction. Like passive tags, the size of semi-passive tags can vary depending on their range and functionality. Because of their batteries, though, semi-passive tags are substantially larger than passive tags. Battery power in semi-passive tags also increases memory capacity, which allows room to implement larger components. These tags can remain in operation for an extended lifetimes, but their batteries need to be monitored and eventually replaced.

Although there is not much demand for semi-passive tags at the moment, manufacturers believe that the use of semi-passive tags will grow to help satisfy the industry requirements that have been mandated by companies like IBM and Wal-Mart.

Sensors can be attached to semi-passive tags to collect environmental data (such as temperature), which is stored in the memory of tags. For instance, semi-passive tags can be attached to a fresh produce case, and the receiving warehouse can check to determine if the food was exposed to any damaging temperatures. Semi-passive tags offer the benefit of using additional power to avoid the read defects caused by distance or material that absorbs or reflects RF. Additional power yields higher signal strength; therefore, these tags have greater potential for wider read ranges. The success and popularity of semi-passive tags will depend on whether RFID users and manufacturers decide to invest in and promote semi-passive tags.

Active Tags

In a general sense, active tags are used for more complex RFID solutions. **Active tags** contain thier own battery source, broadcast their own signal to an interrogator, and therefore do not rely on the interrogator for power. Because of this internal power, active tags achieve the greatest read ranges. Some active RFID tags can send a signal one kilometer.

In some cases, an active tag can be integrated with a GPS (Global Positioning System) to pinpoint the exact location of an item. An active tag's onboard battery can run the tag circuitry all the time or just when a "wake-up" signal is received. A tag operating all of the time will act as a type of beacon, broadcasting its location at specified intervals.

Active tags are the largest because of their batteries. Typical active tags have a thickness of approximately one-half inch and a surface area of 1.5 x 3 inches. Greater tag size also allows for more memory capacity and functionality. The information contained in an active tag can be more complex than that in a passive or semi-passive tag. Active tags can store an object's standard serial number as well as the full list of contents in the container, the container destination, and the origin of the container. With all of this information available, active tags offer up-to-date supply chain information. As with semi-passive tags, active tags' batteries have finite lifetimes and will need to be maintained or replaced accordingly.

The high cost of active tags makes them unattractive for disposable retail product applications; however, active tags can be cost-effective for asset control applications. Cases, pallets, or vehicles can be identified and tracked, which can lead to cost savings that more than offset the cost of the RFID system, including the tags. Currently (mostly because of tag costs), there has not been much of a push for active tags in the RFID industry. Additionally, few domestic and international standards have been established for active tags, which is another hindrance to implementation. In the foreseeable future, active tag solutions will only be used in top-tier RFID systems in the near future.

Identifying tag functionality is a useful way to categorize tags. But as the RFID industry has progressed, other criteria for functionality have been developed on the basis of tag read-and-write capacity.

Classes of Tags

The Auto-ID Center, established by the Massachusetts Institute of Technology to classify tags, proposed five classes of tags, each with greater functionality than the class just below it. These classes define the read-and-write capability of each of the tags in a numerical progression from 0 to 4, with 0 having the least capability and 4 having the most. In addition, each class has the functionality of the class below it. The different classes are Class 0/1, Class 1 Generation 2, Class 3, and Class 4. Generations, such as Class 1 Generation 2, refer to the revisions within each class. (Table 5.1)

Table 5.1 RFID Classes

Class	Functionality
Class 0/ Class 1	Class 0/Class 1 tags are 64-bit read-only passive tags.
Class 2 (Gen 2)	Class 2 tags are passive tags with additional functionality including memory and encryption.
Class 3	Class 3 tags are semi-passive tags that support broadband communication.
Class 4	Class 4 tags are active tags with read-write functionality and greater read ranges.
Class 5	Class 5 tags are essentially interrogators that can power other Class 1, 2, and 3 tags and communicate with other class 4 tags and with each other wirelessly.

Class 0/Class 1

The term “Class 0” was coined to describe a system in which the chips are pre-programmed with a unique code during assembly. Class 0 does not allow RFID tags to be programmed by users in the supply chain. Class 0 tags are passive, UHF-based, and low-cost. Therefore, these tags are ideal for applications that detect only a tag’s presence, not a tag’s data, such as anti-theft devices. Class 0 and Class 1 tags are considered the most basic of the RFID tag classes, because they exhibit the lowest level of operation required by all the classes. For a passive-backscatter tag to fit within the Class 0/Class 1 category, it must have the following:

- An EPC identifier
- A tag identifier (TID)
- A kill function that permanently disables the tag
- Optional password-protected user control
- Optional user memory

Passive, Class 1 tags are UHF- or HF-based. These WORM structured tags are field programmable and still low-cost. Although they are used with applications similar to those that use Class 0 tags, they demand different implementation plans and equipment. Class 1 Generation 2 was created to establish a single, global specification for passive UHF tags.

Class 1 Generation 2

In December 2004, EPCglobal, the organization that took over the Auto-ID operations in 2003, approved the Class 1 Generation 2, air interface protocol in order to establish a single UHF specification. Previously, there were several specifications, including EPC Class 1, EPC Class 0, and two from ISO. Class 1 Generation 2 improves on previous protocols by enabling improvements in:

- Performance of data rates and anti-collision techniques
- Reliability in tag counting
- Performance of interrogators in close proximity
- Security
- Extendibility to higher-function tags and systems

Class 1 Generation 2 protocol is for a full read-write 128-bit tag. In this class, 96 bits are used for the EPC, and the additional 32 bits are for error correction and a kill function. There is still much debate on whether Class 1 Generation 2 tags should have additional memory and encryption. The argument is that these features could be made optional for Class 1 tags, but some people beleive that these features would increase the cost far too much for the tags to be practical.

Class 2

Currently, there is no Class 2 specification in place. Class 2 tags are expected to be read-write, have more memory than Class 1 tags, and support encryption of data on the tag. Class 2 tags are also passive with the following features:

- An extended TID
- Extended optional read-write user memory
- Authenticated access control

Class 3

While Class 1 Generation 2 has become the standard for a majority of the new RFID implementations, Class 3 tags are being used more frequently in more diverse and innovative applications. Class tags are passive/active hybrids (see semi-passive tags described above), which contain a battery that is used as an internal power source for the logic portion of the circuit. Because of these features, Class 3 tags must contain the following:

- An integral power source
- Integrated sensor circuitry

Class 3 tags remain in passive form until they are activated by the interrogator. These semi-passive tags are a viable option for applications that need more functionality than Class 0 and 1 tags, but not as much functionality as Class 4 tags. The Class protocol is a semi-passive system that can be reprogrammed multiple times.

The additional capabilities of the Class 3 tag compared to the lower classes include the ability to identify, communicate, remember, and sense environmental conditions. These abilities create new opportunities for use, including:

- Tracking the location and management of important resources in industrial operations
- Monitoring grocery or pharmaceutical items that require attention to temperature, humidity, and expiration
- Providing a secure entry point for high security zones or providing secure transactions for manufacturing facilities

In supply chain applications, Class 3 tags can be used on pallets and containers to provide an onboard background history by collaborating with the tags on each of the individual units on the pallet.

Class 4

Class 4 tags are active tags, as described in a previous section. The following are Class 4 tag features:

- Tag-to-tag communications
- Active communications
- Ad hoc networking capabilities

Class 4 tags are used in amusement park systems to allow parents to keep track of their children's location. They can also be used on a cargo container to pass on the information from other tags inside the container to an external interrogator. Combined with GPS systems, Class 4 tags can be used to track assets globally.

Tag Placement

One of the biggest challenges of RFID application is tag placement. Unfortunately, there is no clear-cut solution for tag placement. This is primarily because, as the object for placement varies, so does the area on which the tag should be placed. To find the right solution, tests should be done on the item to be tagged. Generally, this requires a process of trial and error on the item itself. Before optimal tag placement can be determined, it is essential to locate seven or eight optimal locations to test for placement. This can be done by using basic knowledge of the laws of electromagnetism or by using an automated RF Visualizer (software that gives a graphical comparison of readability from several tag locations) to show the RF-friendly areas on the item in question.

After choosing the best locations to test and compare, using the RF pyramid method is helpful in determining where and how to tag an item. The RF pyramid method was developed by the RFID Expert

Group (REG) for AIM Global (Figure 5.7). The high point of the pyramid is the most RF-transparent. The higher the RF tranparency is, the more easily a tag can be read through the material. If the item is made up of many different components composed of different materials, each component should be evaluated by its RF transparency using the pyramid.

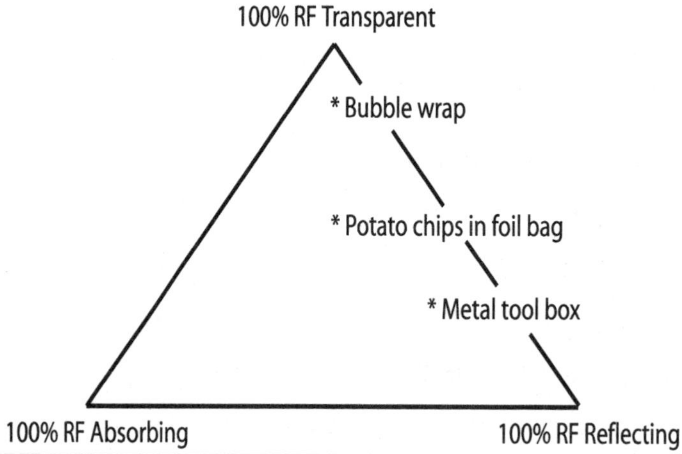

Figure 5.7 RF Pyramid

Much of the time, the tag will go on the outside packaging material, which can be cardboard, plastic, or even metal material. If the material is cardboard or plastic, it is near the top of the pyramid, and therefore RF-transparent. If it is metal, it is not transparent, and a specific made-for-metal tag must be used. One technique used in made-for-metal tags is to put a spacer between the tag and the container. This distances the tag's antenna from the reflective metal, allowing for more normal antenna patterns.

Once the RF transparency of the outside packaging is determined, the next step is to determine the transparency of the contents of the package. If the contents are not RF-transparent, the number of possible locations is limited.

After eliminating a few locations, the remaining locations should be tested with the actual equipment that will be used during normal operation. Testing tags must be done in an RF-friendly location in order to neutralize the effects of the other variables involved in RFID operation. In other words, the tag should be the only component that varies when testing.

In RFID operation, many other factors affect how a tag and interrogator interact. Of specific importance to tag functionality is how RF waves propagate before reaching the intended destination. These factors include:

- Absorption
- Reflection/refraction
- Dielectric effects
- Standing waves and multi-pathing effects

Absorption occurs when some materials absorb energy from the interrogator's antenna, resulting in a loss of available power for sending the signal back to the interrogator. **Reflection** and **refraction** are factors in situations where the material around the tag reflects or refracts the wave. This can result in a distortion of the original signal. **Dielectrics** (electrical insulators) resist the flow of electric current and have a detuning effect on a tag's antenna. **Standing waves** and multi-pathing effects refer to phenomena that cause interference in an RFID system. Standing waves can cancel out the direct waves that are being sent to the tag. The act of multi-pathing is a condition in which the radio wave splits and arrives at the receiver via different paths. When the separate waves arrive at the receiver, they are out of phase and can cancel each other out.

In the interrogation zone, the properties of various materials will cause RF waves to act in specific ways, directly caused by interaction with the material. Table 5.2 charts the effects of numerous materials on RF signals.

It is important to remember that each item is different; there is no clear-cut solution, so trial-and-error testing is vitally important to any RFID system. Testing is essential because optimal location is just as important as a line-of-sight path between the interrogator and the tag. A change in location by even a few centimeters can make a huge difference in tag performance.

Table 5.2 Effects of Materials on RF Signals

Material Composition	Effects on RF Signals
Corrugated cardboard	Absorption (from moisture soaked into the cardboard)
Conductive liquids	Absorption
Glass	Attenuation
Groups of cans	Multiple propagation effects; reflection
Human body/animals	Absorption; detuning; reflection
Metals	Reflection
Plastics	Detuning (dielectric effect)

Summary

The selection and configuration of tags in an RFID system are determined by the power and memory of tags. These aspects of tag function have been categorized into five classes by the Auto-ID center and EPCglobal. In RFID systems, the placement of a tag is crucial to its performance. Trial and error and various tools can be used to test the placement of tags. When designing and configuring a system. It is also important to anticipate how a tag will interact with its environment.

Review

Questions

1. A tag's antenna typically needs to measure _____ of the original frequency's wavelength.
   a. half
   b. 2/3
   c. 25 %
   d. 0%

2. RO tags provide a _____ level of security as RW tags.
   a. lower
   b. higher
   c. similar

3. Which of the following is most RF-absorbent?
   a. Bubble wrap
   b. Plastic bottles of water
   c. Empty pop cans
   d. Metal doors

4. Which of the following tag types use internal battery power (circle all that apply)?
   a. Semi-passive
   b. Active
   c. Passive
   d. Semi-active

5. What will the ¼ wave antenna length be for an RFID tag that operates at 134 kHz?

6. Match the tag type with the correct class:

a. Passive	Class 0
b. Semi-passive	Class 1
c. Active	Class 1 Generation 2
	Class 2
	Class 3
	Class 4

7. What class of tags includes integrated sensors?

   a. Class 0/1
   b. Class 1 Generation 2
   c. Class 2
   d. Class
   e. Class

8. What class of tag would be used to collect information from other tags in the vicinity?

   a. Class 0/1
   b. Class 1 Generation 2
   c. Class 2
   d. Class 3
   e. Class 4

9. Select the best tag placement on a case of bottled water from the following choices: (see figure in file "tags on cases").

10. Select the best tag placement on a book that will be read both on the shelves and as it is "checked out."

    a. In the middle of inside front cover
    b. On the top left side of the inside front cover
    c. In the middle of inside back cover
    d. On the spine

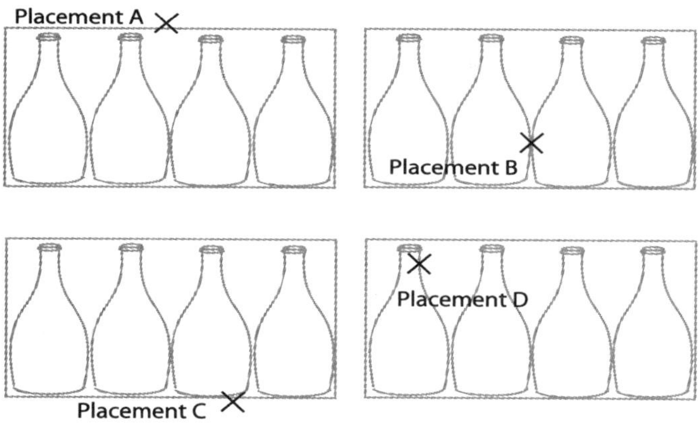

11. Which of the following is the least RF absorptive?

    a. Paper
    b. Glass
    c. Human body
    d. Drinking water

Discussion

1. Describe the technique used to determine tag placement on a new item.

2. Describe the difference between RF absorbing and RF reflecting materials. What affect will the presence of these materials have on tag placement?

3. Discuss how the fundamental principles of circuits and antenna affect the functionality of tags.

4. Explain how various environmental conditions and materials will affect the communication of tags. Name some of the specific materials that affect radio signals, and describe these effects.

5. Describe the use of the RF pyramid to determine tag placement. After the pyramid process has been conducted, how should tag placement be decided?

6. Describe the difference between RF absorbing and RF reflecting materials. What effect will the presence of these materials have on tag placement?

Chapter 6
Interrogators

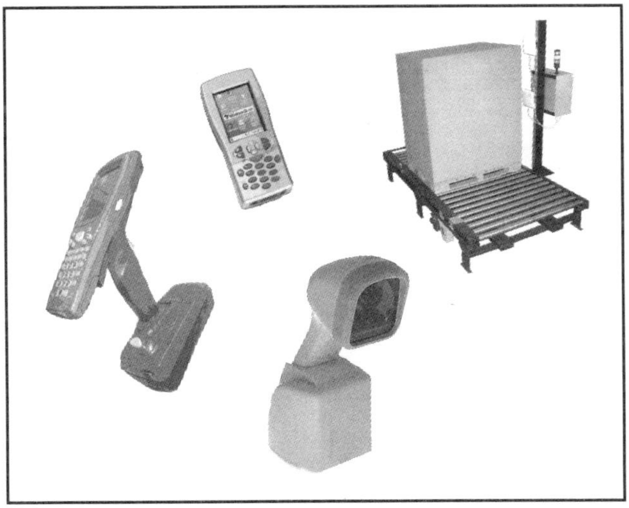

Chapter 6: Interrogators

Key Concepts

- Interrogator types, function, and operation
- Interrogation zones
- Applications software
- Physical setup
- Dense interrogation

Terms

- Antenna radiation pattern
- Application layer software
- Bi-static antenna
- Dense interrogator mode
- Edgeware
- Firmware
- Fixed-position interrogator
- Full duplex
- Half duplex
- Handheld interrogator
- Interrogation zone
- Interrogator/encoder
- Middleware
- Mobile mount interrogator
- Mono-static antenna

This chapter covers interrogator function, interrogator types, physical setup issues, and software setup. The interrogator, also called a reader, has the largest workload of all the components in an RFID system. It is responsible for activating the tag, receiving information from the tag, and relaying that information to the data collection system (Figure 6.1). The interrogator creates and amplifies the RF signal that is sent out to an antenna, receives the tag's response signal from the antenna, amplifies and demodulates the signal, organizes the data from the tag, and either stores it or sends it to a computer or network.

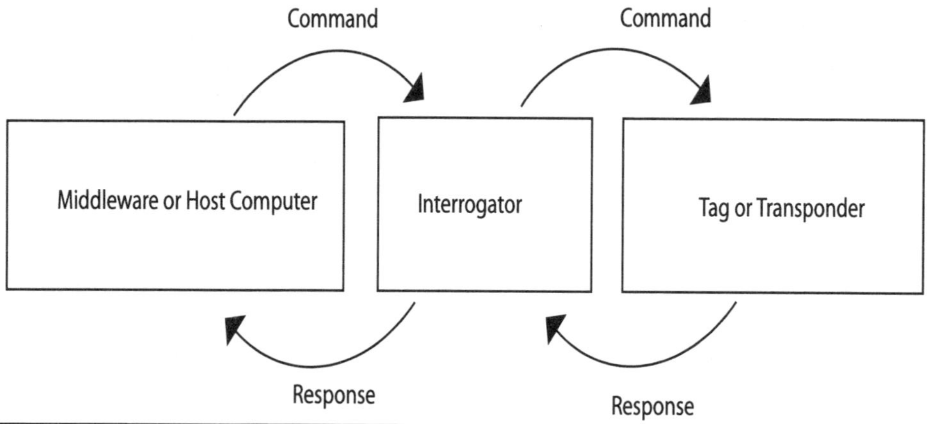

Figure 6.1 Interrogator Responsibilities

Interrogator Communication

In order for the information from the tags to be useful, the interrogator must communicate it to a person who can use it. This means that all interrogators must either have some type of display (as in handheld interrogators) or interface with a computer display. If the interrogator uses a computer interface, a wireless or cable connection between the interrogator and computer is required.

The simplest type of connection is a serial (typically RS232) connection, in which a computer is connected with a serial cable to the interrogator. Besides having the proper connectors, it is also important to utilize the right cabling configuration. A "null modem" cable crosses the transmit and receive lines, while a "straight" cable assumes the computer will correctly direct the signal received from the interrogator (Figure 6.2). Application software must be loaded and executed on the computer to enable communication with the interrogator.

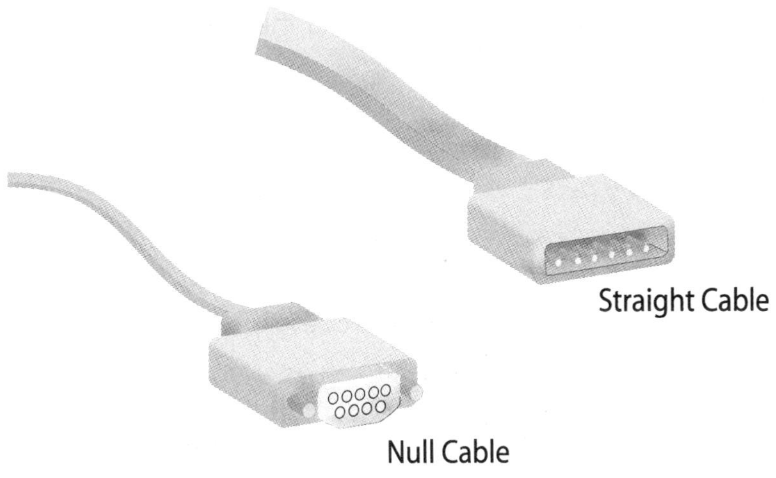

Figure 6.2 Null Cable

Instead of connecting directly to a computer application, an interrogator may utilize a network protocol (such as Ethernet) to communicate with a network hub. In this case, the interrogator must be assigned an IP address unique to the network. Edgeware software executes on the interrogator and allows computers on the network to collect tag information and control the interrogator's function.

For installations where the interrogator is mobile, or where running network cable is impractical, a wireless network interface (Wi-Fi) can be used. The edgeware that operates on the interrogator is the same, but the physical layer interface is done through a wireless networking protocol such as IEEE 802. Although many working installations successfully utilize wireless networking, the wireless network may operate on the same frequency as the RFID system and affect the interrogation zone performance.

Interrogator Types

Interrogators come in a variety of sizes, shapes, and models. This section will explain some of the basic physical options available. Hardware and software specifications should be considered when choosing an interrogator for an RFID system. For RFID+, a technologist must identify the various interrogator types and understand the differences in functionality among these types. The functionality of an interrogator determines its appropriate configuration in an RFID system.

The size of an interrogator and the number of its functions will determine the overall cost of the hardware. Each type of interrogator is designed for specific uses. The following are the most common types:

- Handheld
- Fixed position
- Mobile mount
- Interrogator/encoder

Handheld interrogators present maximum flexibility and are typically small and durable (Figure 6.3). They are quite practical for general use because they allow the user a certain amount of mobility. A user can bring the interrogator to a desired tagged object and gather all of the information.

Fixed-position interrogators (Figure 6.4) are placed at specific locations, such as conveyor belts, doorways, or loading docks. They also can be placed in a portal configuration (Figure 6.5) to maximize read accuracy and create an easy read point. Forklifts can drive through a portal configuration as they load a truck with no wait time. The loading process can continue while the information is being updated in the main data system.

Figure 6.3 Handheld Interrogator

Figure 6.4 Fixed-Position Interrogator

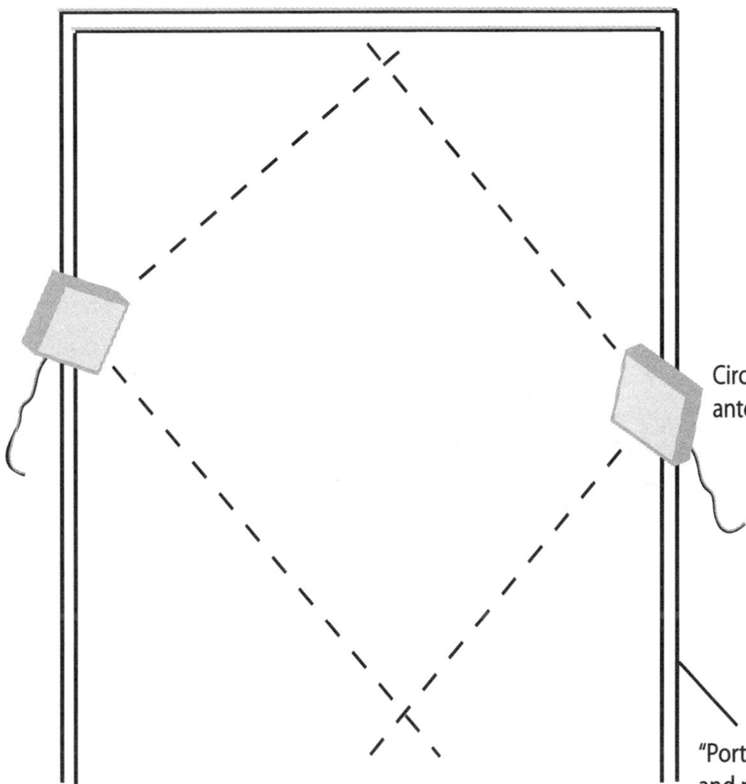

Figure 6.5 Portal Configuration

Mobile Mount Interrogator

The **mobile mount interrogator** can be placed directly on a vehicle (Figure 6.6). A forklift, for example, can simply drive down the aisles of a storage room as the interrogator transmits and receives information from transponders on pallets or other objects. This is a very efficient method of tracking cases, pallets, or other goods.

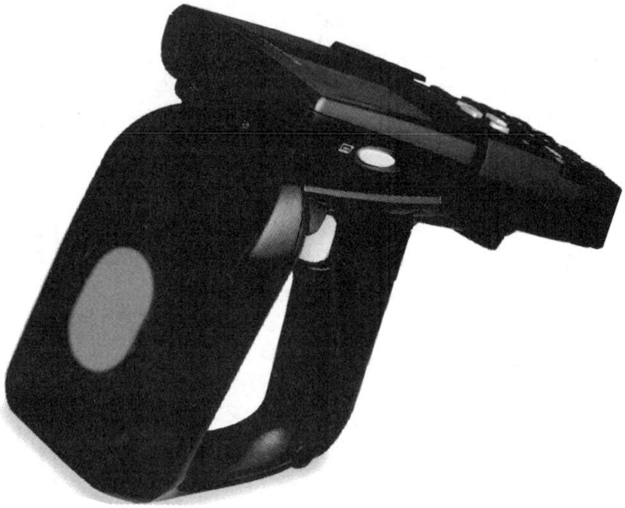

Figure 6.6 Mobile Mount Interrogator

Interrogator/Encoder

The **interrogator/encoder**, also called a read/write interrogator, can encode data into the tag and interrogate this data. If the end user is willing to pay the cost, any RFID interrogator can be modified to operate as an interrogator/encoder. In most situations, a read-only interrogator is necessary. The interrogator/encoder will usually look just like a read-only interrogator because the only difference is the functionality of the interrogator.

> Exam Note
>
> All interrogators can send out a signal to wake up the tag and receive the tag's information.
>
> Not all interrogators can wake up a tag, write information to it, and receive information from it.

Interrogator Specifications

There are some other physical specifications to be considered when choosing an RFID interrogator for a particular system. Most of this information can be found on the data sheet of a specific vendor's interrogator model. The following are the most important things to look for:

- Operating frequency
- Supported RFID tag protocols
- RF output power
- Number of antennas that can be used with it
- Interrogator communication protocol
- Configuration software availability
- Upgrade availability (future use)

These specifications help define which interrogator is best suited for the RFID application. It is crucially important to understand why the RFID system is being installed and what the end user hopes to accomplish with RFID implementation. The design of a successful RFID system depends on matching the end user's specifications with the right components.

Half and Full Duplex

Interrogators can operate in two modes: **half duplex** and **full duplex**. Duplex specifications are determined by the way data transmits through a medium, which depends on the capability of the individual interrogator. Half duplex (HDX) means that only one part of the data transmission can be accomplished at a time. An example of a half-duplex system would be walkie-talkie communication. Only one side of the system can talk at a time.

A full-duplex (FDX) interrogator can send and receive at the same time. An example of full duplex communication is two people talking at the same time in a room or on the phone. Unlike humans, the interrogator can transmit legitimate data and understand incoming data at the same time. In order for an interrogator to operate in a full-duplex mode, it must utilize **bi-static antennas**. A bi-static antenna configuration uses two antennas: one to transmit and one to receive. If an interrogator uses **mono-static antennas** (one antenna both transmits and receives), it can only operate in half-duplex mode.

Software Configuration

Along with the hardware and configuration setting for interrogators, a technologist should be familiar with the software the interrogator uses as well as the software used by the components with which the interrogator operates. There are several layers of software used in RFID systems. These layers may all be individual programs or may be combined into one application. As vendors offer more complex solutions, the lines between layers blur. In this discussion, the layers will be explained separately; but software packages encountered in the field may encompass one, two, three, or all four layers.

The lowest level of software is called **firmware** and is stored inside the interrogator in some form of Read-Only-Memory. Firmware is the program for the interrogator's microprocessor, which interfaces with the very lowest level of hardware. Typically, firmware is stored in an EEPROM that can be updated in the field. Vendors distribute firmware updates electronically (by e-mail, Web sites, or FTP) or send notifications when updates are available. Instructions with the update will explain the necessary hardware connections between the interrogator and the host computer, and the installation procedures.

Edgeware

Edgeware is the software that runs on the host computer that interfaces with the interrogator. With standalone interrogator systems that do not require a host computer to operate, edgeware runs on the interrogator. It allows configuration of the interrogator's functions, including the read rate, power level, frequency of operation, and method to poll tags. Edgeware also serves as a data conduit between the bits collected by the firmware and the data collection software. The edgeware may organize some information by consolidating multiple reads from one tag or eliminating responses from item-level tags when only case tag information is desired.

Middleware

Middleware is responsible for the data collection, management, and flow among the tags, interrogators, and inventory applications. It plays an important role in the business justification for RFID applications. It communicates the data collected by the interrogators to the **application layer software**, such as inventory management software used to make business decisions. Without this interface, the data collected by RFID installations cannot be easily integrated into the decision-making process. Middleware is often written by a different vendor than the interrogator and the application software vendor, but can be customized to each customer's requirements.

Firmware and edgeware are within the scope of a technologist's responsibilities, while middleware and application layer software are within the purview of the customer's information technology group. The RFID technologist may take part in testing of the interface between the edgeware and middleware. Some technologists, based on their employers' policies, may also have some responsibilities for the middleware; however, RFID+ certification does not focus on middleware or application layer software.

Interrogation Zone

The ultimate purpose of most RFID systems is to provide effective tracking and information gathering for objects at the lowest cost. A primary way that RFID users ensure their systems are cost-effective is to optimize the **interrogation zone** of their system. The interrogation zone is an area defined by the antenna(s) in which the tag can be energized and spoken to by the interrogator. This zone can be adjusted by relocating the antenna and/or adjusting the function of the interrogator. There are several things to consider when adjusting interrogation zones, such as output power, antenna direction, extraneous noise, and antenna footprints. Proper on-site analysis can address most of these considerations.

Configuring the Interrogation Zone

Antenna radiation patterns (Figure 6.7) determine the outline of the interrogation zone. The combination of the number, type, and direction of antennas determines the three-dimensional shape of the RF propagations from the antennas and the area where signals from tags can be received.

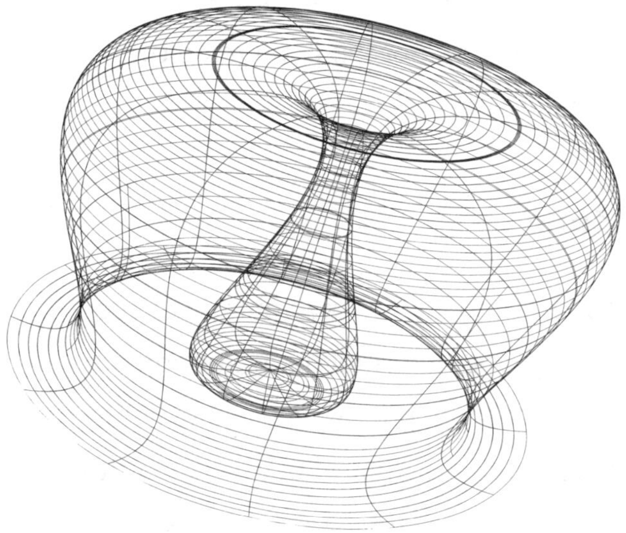

Figure 6.7 Antenna Radiation Patterns

Initial preparation involves mapping the antenna's footprint. The interrogator and antenna specifications will give a good indication of the theoretical shape. Mapping involves taking a tag or tags and placing them at certain points away from the antenna to test the read rate. This process should be done for the entire antenna radiation to verify that the antenna is optimized with the interrogator.

The number of antennas used determines the size and shape of the interrogation zone. A simple, low tag count zone may be established with only one antenna. In this case, the zone and the antenna pattern are the same. Better definition or larger zones require more antennas.

A door zone can be created with two antennas tied to the same interrogator (Figure 6.8). The pair of antennas can cover more volume than a single antenna, and ensure coverage in the center of the passageway where tags are most likely to pass.

Four antennas can create a zone around a conveyor: an antenna on either side, one above, and one below the conveyor (Figure 6.9). This placement ensures that tags will be read regardless of the orientation of the box coming down the conveyor. Tags on the leading or lagging edge will be read as the box is entering or leaving the zone, and tags on the other four sides will be read by the antenna they face. One interrogator powers all four antennas. Either plastic rollers or a plastic belt works most effectively as the conveyor mechanism inside the zone. If the conveyor were made of metal, the reflections would block signals from the bottom antenna and scatter signals from all the others.

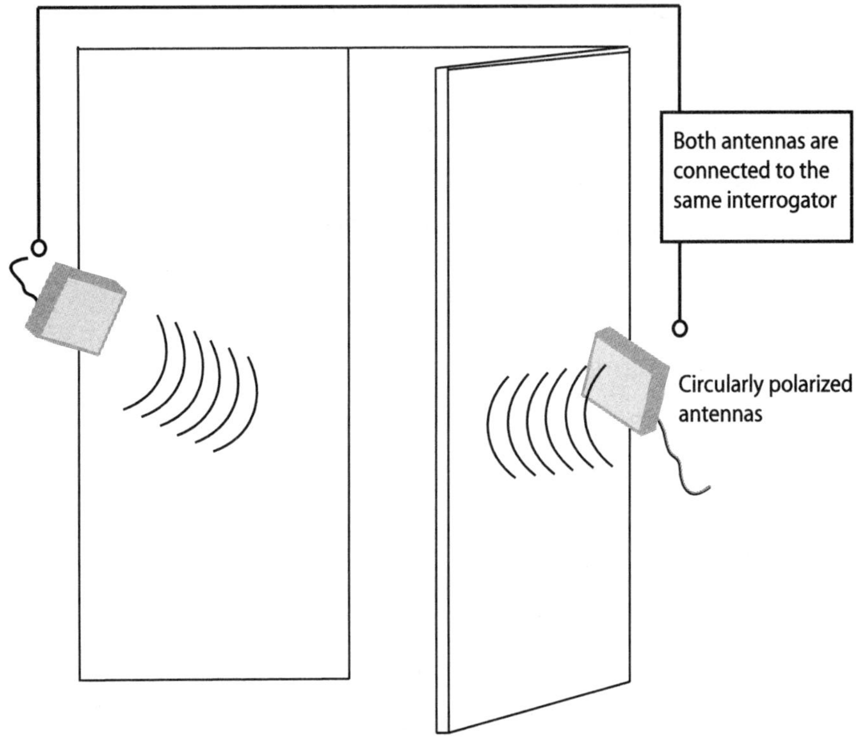

Figure 6.8 Door Zone

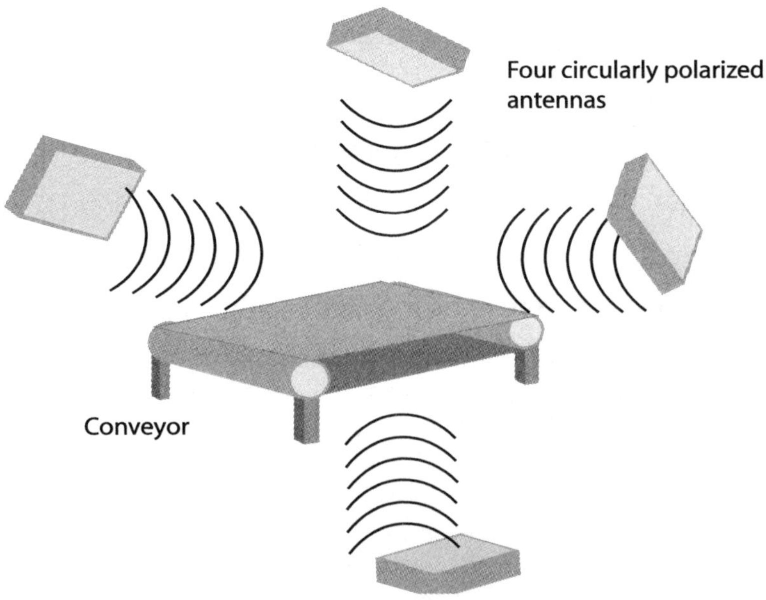

Figure 6.9 Conveyor Zone

The range of an interrogation zone can be augmented by increasing the interrogator output power. Increasing the power increases the read rates and makes the zone less sensitive to materials that reflect or absorb RF; however, there are a couple of potential problems caused by simply increasing the interrogator's output. Output power is regulated by individual countries. For example, in the United States, the FCC regulates radio communications. An increase in the output power to broaden the interrogation zone range can have serious repercussions. It could be illegal or result in the jamming of other RF systems.

The zone can also be contained with RF reflective material, such as a metal sheet or mesh. This will keep the RF signal in the zone by reflecting the signal back toward the target rather than losing the RF energy to free space.

Dense Interrogator Mode

Two interrogators working in proximity (such as neighboring dock doors) can interfere with each other. Two or more interrogators may read one tag. The signal from interrogator B may swamp the signal coming from a tag inside interrogator A's zone. This may block the signal entirely, or may just slow down interrogator A's singulation or read rate. The first line of defense against such problems is to tune each interrogation zone properly so that zones do not overlap.

Another way to prevent interference between interrogators is in the judicial use of frequency slots. **Dense interrogator mode** is an option for interrogators certified for Generation 2 operation. It is specifically designed for environments where there are many interrogators. Besides employing frequency hopping as in Generation 1, the interrogator signal that powers the tags is in the center of the channel, and the tags respond on the two sides of the channel. Since interrogator signals are separated spectrally from the tag signals, one interrogator cannot interfere with a tag response to another interrogator. Figure 6.10 below shows the spectrum requirements in the United States. In Europe, tags and interrogators are assigned alternating channels.

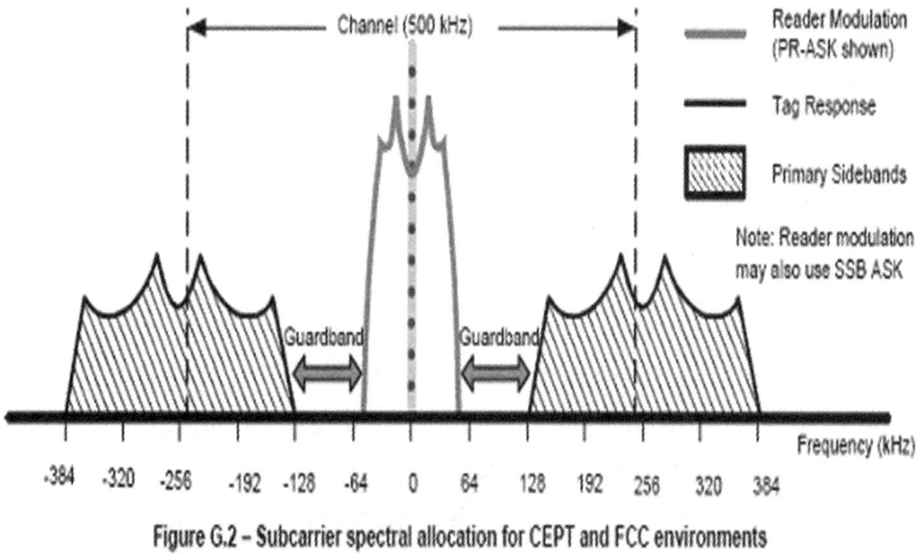

Figure 6.10 United States Spectral Requirements

In addition to interrogators interfering with each other's RF operation, data interrogators may load down the network infrastructure and interfere with network operations. The technologist should verify that the customer's network administrator has adequate bandwidth available for the traffic generated by any interrogators installed.

Anti-Collision Protocols

In situations when there are multiple tags in the interrogation zone, some type of anti-collision protocol must be used. Since the interrogator can only process a response from one tag at a time, it is important to singulate the tags' replies to the interrogator's inquiries. The interrogator must command which tag should respond at any instance in time. Various protocol types are used to accomplish this task; the three most common will be discussed. Both the interrogator and tag must be designed or programmed for the same scheme.

The Aloha protocol relies on the tags in the read zone to vary the time at which they respond. After a wake-up call from the interrogator, the tags wait to respond; the time they wait may either be generated randomly inside the tag, or it may be assigned by the interrogator. Once the interrogator has received a specific tag's information, the interrogator commands that tag to sleep. If the protocol is based on each tag's random response, eventually all the tags will respond alone and be read one at a time, but a tag may have to respond more than once if it responds at the same time as another tag. If the interrogator assigns time slots to individual tags, they will have to wait their turn to reply, but will only have to respond once.

The binary tree or "tree-walking" protocol uses EPC numbers to determine the order in which the tags will be read (Figure 6.11). The reader wakes up all the tags in the field, then asks all the tags with a "1" in the most significant bit to respond, the others to sleep. If there is more than one tag that responds, the interrogator asks all tags with a "1" in both the MSB and the second significant bit to respond. The interrogator keeps "walking" down the tree until only one tag responds. Then it "walks" back up the tree to previous level to address the tag with a zero in the last bit. The interrogator keeps going up and down the tree until all the possible numbers have been addressed for which there is a possibility of tag response.

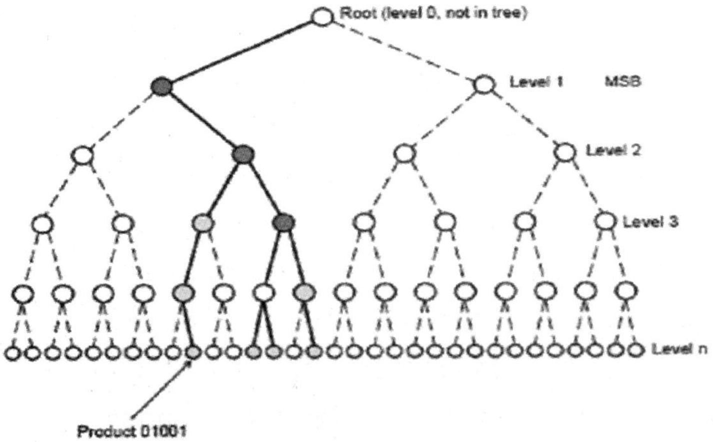

Figure 6.11 Protocol Specifications

The protocol used in Gen 2 tags is the "Q" protocol. It is possible for more than one tag to have the same EPC number. The "Q" protocol allows for this, where the binary tree method does not. The interrogator wakes up the tags in the zone, and each tag responds with a random number. It is not unlikely for two tags to reply with the same number, thus each number should be unique, even if the EPC numbers are the same. The interrogator then addresses each tag by its random number and receives its information. After the tag is read, it goes into an alternate state. It is not sleeping in this state, but knows that it is not to reply again without another wake-up call from the interrogator. This saves time when tags might have to be cycled several times to ensure that all the tags in the zone have been read.

Summary

The interrogator activates a tag, receives information from a tag, and communicates this information with a host computer or other network device. There are many types of interrogators used for RFID applications, and they are designed for their physical and technical functionality. The range of an interrogator's signal, the interrogation zone, can be modified using the interrogator's configuration and antenna. An RFID application may use more than one interrogator, but the site will have to be set up accordingly.

Review

Questions

1. Which of the following should not be used for the connection between the interrogator and host computer?
   a. Ethernet
   b. IEEE 802 wireless
   c. ISO 18000
   d. RS232 null modem

2. What primary type of interrogator would be utilized at an incoming dock door?
   a. Handheld
   b. Fixed position
   c. Mobile mount
   d. Interrogator/encoder

3. What secondary type of interrogator would be utilized at a dock door?
   a. Handheld
   b. Fixed position
   c. Mobile mount
   d. Interrogator/encoder

4. An interrogator uses mono-static antennas. This means that _____.

   a. it can only operate at one frequency and it does not support frequency hopping

   b. it can only operate in full duplex mode

   c. it can only operate in half duplex mode

   d. it can operate in full- or half duplex mode

5. True of False: Firmware should typically be edited by the technologist to adapt the interrogator for each installation.

6. Which of the following is typically not done by edgeware?

   a. Consolidates multiple reads from one tag

   b. Configures power level of the interrogator

   c. Configures polling method

   d. Interfaces with inventory management software

7. Name three considerations when installing multiple interrogators in close proximity to one another.

8. A null modem cable _____.

   a. is the same as a coaxial cable

   b. crosses the transmit and receive lines

   c. is always a fiber-optic cable

   d. is not used with interrogator components

9. When an interrogator(s) is connected to a network protocol, it works with the _____ software to communicate information to the network.

   a. firmware
   b. middleware
   c. edgeware
   d. host computer

10. Which of the following specifications can usually be found on a vendor's interrogator data sheet?

   a. Configuration software availability
   b. Available upgrades
   c. Supported tag protocols
   d. All of the above

11. The _____ is affixed to a vehicle and scan pallet tags.

   a. mobile-mount interrogator
   b. handheld interrogator
   c. dense interrogator
   d. fixed-position interrogator

12. Middleware is responsible for _______.

   a. the configuration of an interrogator's function
   b. the operation of an interrogator's microprocessor
   c. the graphic presentation of data on the interrogator's GUI
   d. the collection, management, and flow of data

Discussion

1. Suggest an antenna location for an interrogator mounted on a fork lift. Describe the advantages and disadvantages of your selection. What type of connection would be utilized to communicate with the middleware application?

2. The interrogation zone has been considered in other chapters. Discuss how interrogators can be selected and configured to achieve specific interrogation zone conditions.

3. Explain antenna footprints. How are footprints mapped, and what do footprints indicate about the interrogation zone?

4. What should be a technologist's considerations for configuring multiple interrogators on the same site?

5. One way to increase read rates is to increase the power out of the interrogator. What are some possible consequences of this action?

Chapter 7
Regulations AND Standards

Chapter 7: Regulations and Standards

Key Concepts

- RFID regulation
- ISO standardization process
- EPC and EPC classes
- FCC RFID standards
- Data security

Terms

Authentication protocols

Cryptographic key

Electronic Product Code (EPC)

EPCglobal

Federal Communications Commission (FCC)

International Organization for Standardization (ISO)

Object Name Server (ONS)

Physical Markup Language (PML)

In the RFID industry, the absence of a global organization to govern RFID regulations has been problematic and has prompted a number of countries to establish their own regulations. In response to this situation and in an attempt to ensure broadly recognized standardizations, a handful of institutions have established RFID regulations to make RFID systems secure and safe. Additionally, standards can benefit companies by prompting technological advancement, reducing costs, easing implementation, and improving quality and confidence in RFID technology. This chapter will survey the standards that have been established by the International Organization for Standardization (ISO), the Federal Communications Commission (FCC), EPCglobal, and several countries that have launched independent regulations.

A summary of regulations is helpful in determining which standards apply to specific systems, components, and protocols. During the design and implementation process of any RFID system, a technologist must recognize the regulations that will apply to the system.

It is important to recognize that the information on standards provided in this chapter is not comprehensive, and is intended as a broad reference. A technologist should reference complete guidelines and standards that apply to the requirements and specifications of the system with which he or she is working.

The International Organization for Standardization (ISO)

The **International Organization for Standardization (ISO)** attempts to establish standards for RFID by the consensus of manufacturers, users, consumer groups, research laboratories, governments, engineering professions, and academic organizations. It is a global organization, and all member countries agree to follow its standards. No local ordinances should contradict any ISO standard. Recent committees have established standards for agricultural electronics, RFID for item management, contact-less integrated circuit cards, supply chain applications of RFID, and other topics relating to RFID. Due to the number and depth of ISO standards, this chapter will provide only a brief overview. The comprehensive standards can be obtained though the ISO's Web site, http://www.iso.org/iso/en/ISOOnline.frontpage.

In an attempt to produce a coherent set of terms and definitions unique to RFID, particularly data capture and automatic identification, ISO established a glossary of terms, ISO/IEC 19762, which is intended for both RFID specialists and nonspecialist RFID users.

Frequency Ranges

A primary feature of ISO's standardization is the frequency ranges it approves. Together the various ISO guidelines authorize systems operating at the following frequency ranges:

- Below 135 kHz
- 13.56 MHz
- 860–960 MHz
- 2.45 GHz
- 5.8 GHz

ISO 18000 Air Interface Standards

The ISO 18000 series standards apply primarily to air interface protocol (Table 7.1). In a simple sense, air interface protocol is the set of rules by which RFID tags and interrogators communicate. The ISO air interface standards outline the modulation, bit encoding, and read/write specifications for the data transmission between interrogators and transponders. They also cover the anti-collision algorithms used in interrogator zones with multiple tags. ISO 18000 standards parts 1, 2, 3, and 4 establish the parameters for the air interfaces of all major frequencies, including the interfaces for tags operating at 135 kHz, 13.56 MHz, and 2.45 GHz. Part 6 of the ISO 18000 standards specifies the air interface for RFID tags operating at ultrahigh frequency (860–930 MHz). The following are the air interface protocol frequencies approved by ISO:

- Below 135 kHz
- 13.56 MHz
- 2.45 GHz
- 860–960 MHz
- 433 MHz (active)

In each of these frequencies, the parameters of system operation and performance are directly associated with the signal structures of tag and interrogator communication. Specific types of tags, interrogators, and antennas are designed for each of these frequencies.

Table 7.1 ISO 18000 RFID Air Interface Standards

Standard	RFID Category	Topics Addressed	Status
18000-1	General rules for internationally accepted air interface protocol frequencies	• Parameter definition for air interface at any frequency • Methods of avoiding interference with other RFID systems • Methods of allowing interoperability • Migration between different frequencies with minimal effort	Published
18000-2	Air interface below 135 kHz	• Specification of tag types A(FDX) and B(HDX) • Physical layers • Anti-collision protocol	Published
18000-3	Air interface at 13.56 MHz	• Specification for physical layers, collision management, and communication protocol at 13.56 MHz • Two modes, which are not interoperable or interfering • Reader has to support either Mode 1 or Mode 2, or both	Published

18000-4	Parameters for air interface at 2.45 GHz	• Passive and Active modes • Mode 1: Reader talks first • Mode 2: Tag talks first, long-range high-data rate • Anti-collision protocol	Published
18000-6	Parameters for air interface at 860-960 MHz	• Type A and Type B • Encoding, modulation, data rate, etc. • Reader Specifications • Tag Specifications • Frequency Range	Published
18000-7	Parameters for air interface at 433 MHz	• Typical applications operate at ranges greater than one meter • Read/write • Anti-collision	Published

Data Objects and Identifiers

The ISO 15961 standard defines how data objects are presented in RFID systems. Specifically, this standard identifies how data should be transferred among the RFID tag, interrogator, and middleware applications. ISO 15961 guidelines clarify and amend the ISO 8825 encoding rules: 88212-1 Standards for ASN.1 Processes, and 9834-1 Procedures for ASN.1 Object Identifier Trees. Similarly, ISO 15962 addresses the appropriate encoded structure of object identifiers and the compaction rules for this data. ISO 15961 also specifies how applications use and communicate data, including how data is formatted and stored. ISO 15963 defines the numbering system used for the identification of tags. Each tag has a unique identification number to ensure its traceability in a system.

RFID in the Supply Chain

Many of ISO's decisions outline standards for the use of RFID in the supply chain, including vendor, intermediary, and end-user specifications. ISO standards 17364–17367 regulate RFID in the supply chain. ISO 17364 defines supply chain applications for transportation units, primarily freight containers. ISO 17365 governs returnable transport items in the supply chain. ISO 17366 and 17367 define how products are packaged and tagged. ISO 17363–17368 standards cover the different types of logistics containers and packaging used with RFID.

Cards

ISO 14443, parts 1 through 4, specifies the rules for contact-less integrated circuit cards, including physical characteristics, power and signal interface, bit rate, initialization and anti-collision, and transmission protocol. ISO 10536 series standards define the nature and characteristics of the power and communication for close-coupled cards. ISO 10373 covers the methods of testing contact-less integrated circuit cards.

Agricultural Electronics

ISO standards 11784 and 11785 detail the identification of animals with RFID. The ISO 11784 standard primarily specifies the structure of RFID coding used with animal identification systems. ISO 11785 covers the transmission protocols between transponder and receiver. Table 7.2 identifies the number, category, topics, and status of the various ISO standards for RFID.

Table 7.2 ISO Standards

Standard	Category	Topics Addressed	Status
10373	Identification cards: test methods	• Close-coupling cards according to ISO 10536 • Proximity cards • Vicinity cards	Published
10374	Freight containers: automatic identification	• Container identification systems • Data coding systems • Data description • Performance criteria • Security features	Published
10536	Close-coupled cards	• Physical characteristics • Dimensions and location of coupling areas • Electronic signals and reset procedures • Answer to reset and transmission protocols	Published
11784	Identification of animals: code structure published	• Frequency • Baud rate • Bit coding • Data structure	Published

11785	Identification of animals: technical concepts	• Transponder activation • Information storage • Transceiver characteristics • Interchanging methods	Published
14443	Proximity cards	• Physical characteristics • RF power and signal interface • Initialization and anti-collision • Transmission protocol	Published
14443	Contact-Less smart cards	• Operating frequency • Read/write range • Speed • Security	Published
14443-1	Contact-Less Smart Cards: physical charac-teristics	• Card dimensions • Surface quality for printing • Mechanical resistance • UV and X-ray resistance • Magnetic fields	Published

14443-2	Contact-Less Smart Cards: RF power and signal interface	• RF power • Signal Interface • Type A signaling schemes • Type B signaling schemes	Published
14443-3	Contact-Less Smart Cards: initialization and anti-collision	• Polling for PICCs • Byte format • Request and answer to request commands • Anti-collision methods	Published
14443-4	Contact-less Smart Cards: Transmission Protocols	• Half-duplex block transmission protocol • Type A high-level transmission protocols • Type B high-level transmission protocols • Baud rate	Published
15693	Vicinity Cards	• Physical characteristics • RF power and signal interface • Initialization and anti-collision • Extended command set and security features	Published

15961	RFID for Item Management: Protocols for Data Handling	• Interface with applications Systems • Presentation of data objects • Structure of object identifiers • Data transfer commands • ASN.1 processes • Transfer syntax	Published
15962	RFID for Item management: data transmission protocols	• Data processing • Structure of object identi-fiers • Data compaction • Data formatting • Application commands	Published
15963	Unique identification of RF tag and registration authority	• Number system • Procedural standards • Use of unique identifica-tion of RF tags in the integrated circuit	Published
17358	Supply chain applications	Standards are under development	Draft
17363	Application requirements	Standards are under development	Draft

17364	Transport units	Standards are under development	Draft
17365	Returnable transport items	Standards are under development	Draft
17366	Product packaging	Standards are under development	Draft
17367	Product tagging	Standards are under development	Draft
18047	Device conformance test methods	Standards are under development	Parts 3 and 4 Published
18047-2	Device conformance test methods: air interface communications below 135 kHz	Standards are under development	Draft
18047-3	Device conformance test methods: air interface communications at 13.56 Hz	• Mode-specific Conformance parameters • System functionality and interoperability	Published
18047-4	Device conformance test methods: air interface communications at 2.45 GHz	• Mode-specific conformance Parameters • System Functionality and Interoperability	Published
18047-6	Device conformance test methods: air interface communications at 860 to 960 MHz	Standards are under development	Draft

18047-7	Device conformance test methods—air interface communications at 433 MHz	Standards are under development	Draft
19763-3	Framework for meta-model Interoperability	Standards are under development	Draft
24710	Item management	• Elementary tags • Implementation of ISO 18000 series	Published

Electronic Product Code (EPC)

Electronic Product Code (EPC) was developed to connect physical objects to an immense network of computer data electronically. EPC is the numbering scheme associated with a particular tag in the RFID system and is linked to data communications and storage protocols. The EPC number is four sets of binary numbers that link the EPC to its product. In a sense, EPC is the next generation of the barcode (UPC/Universal Product Code). For the moment, EPC will help businesses bridge the gap between existing barcode systems and new RFID systems. Retail supply chains stand to benefit the most from EPC because they need some way to serialize their products. Products move from country to country or business to business, so a standard format needed to be implemented; EPC was born out of this need.

EPC was originally developed by the Auto-ID Center at Massachusetts Institute of Technology (MIT) in 2000. After the Auto-ID Center ceased its efforts on the EPC project, the Uniform Code Council (UCC) and EAN International formed the EPCglobal organization to continue the project. EPCglobal standards are ever changing and current standards are best referenced at the EPCglobal Web site, http://www.epcglobalinc.org.

The EPCglobal Network

EPCglobal is responsible for EPC numbering standards and protocol. The EPC network involves five elements: the actual EPC number, the RFID components (tags and readers), the Object Name Server (ONS), the Physical Markup Language (PML), and Savant Software. The EPC is a series of numbers that identifies a product or object. When an interrogator reads a tag, middleware can communicate information from the tag to the RFID information system.

Technically, the information system is called the **Object Name Server (ONS)** or local application system. The ONS tells the system components where to locate network data regarding the object with the EPC. ONS, for instance, tells the information system when the product was manufactured.

PML (Physical Markup Language) is a language commonly used to define data on physical objects in the RFID system. Savant is a software technology that manages all of the RFID data in the system. It moves and manipulates data so that existing corporate data structures are not heavily congested. If the EPCglobal network were thought of as an organism, Savant would be the central nervous system.

The EPC number does not contain any information about the product; it is just an identification number. The information about the item, which can be of real value in practice, is stored on the ONS and accessed using PML and Savant software.

EPC Format

EPC is broken down into four distinct sections of hex numbers (Figure 7.1). The header, EPC manager, and object class numbers all act like the old UPC numbering system. An optional filter section is also available between the header and the EPC manager.

Each section of the EPC should be understood individually. The first section, the header, is composed of 8 bits, 0 through 7, in a 96-bit EPC structure. The header indicates to the reader what numbers of the code will follow. For example, the header can designate if a military or FDA series of numbers is to follow. The header can also be referred to as the version number. This value specifies length, type, and structure of the EPC.

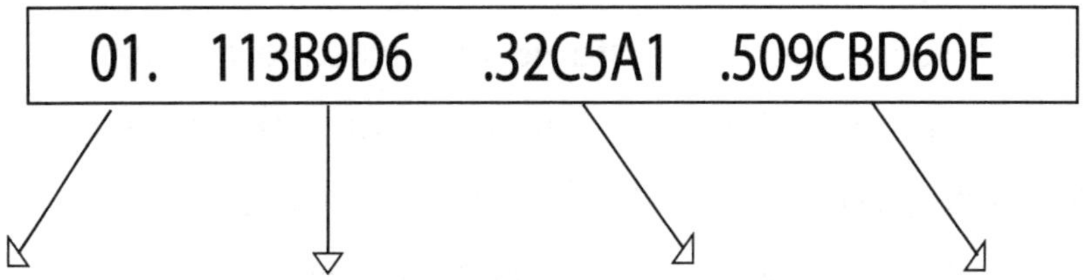

01	1139D6	32CA1	509CBD60E
8 Bits	28 Bits	24 Bits	6
Version of EPC	Manufacturer Identifier	Product Indentifier	Item Serial Number

Figure 7.1 Electronic Product Code

The EPC manager identifies the company associated with the EPC. It can tell who is responsible for a specific set of EPC numbers. EPCglobal assigns the general manager number to an entity upon entry into the EPC Network.

The object class number refers to the exact type of product being tracked. It is very similar to the Stock Keeping Unit (SKU) of a product. The object class number must match up to the EPC manager domain.

A filter value can be implemented if the technologist finds it is necessary for a particular system. It may increase efficiency or enable a more effective read from the RFID tag. The latest advance by EPCglobal is represented by the last set of numbers in the EPC, which represents the serial number of the product. It allows tracking systems within the supply chain to link a tagged product directly to all of its unique information. Barcodes track product types generically, but EPC tracks products individually. For example, the barcode on a box of cereal is a general number that is the same for all other boxes of the same brand. If cereal boxes were tracked with RFID tags, each box of cereal would possess its own EPC identity.

> **Exam Note**
>
> An object class number includes an SKU of some consumer good where the EPC manager number value indicates the company that produced the good.

	Filter Value			
Header	Filter	EPC Manager	Object Class	Serial Number

Figure 7.2 EPC with Filter

The EPC numbers are associated with different types of tags. The size of the tag data defines the range of information available. There are two choices: the 64-bit scheme and the 96-bit scheme. The 96-bit scheme allows more data in applications where 64-bits are not enough (Table 7.3). Mathematically, a 96-bit tag can accommodate as many as 268 million companies at 16 million classes each with 68 billion serial numbers in each class (*RFID Field Guide*, pg. 74). Those numbers show that there is plenty of room for the Electronic Product Code in the future.

Table 7.3 EPC Bit Allocations

	Header	EPC Manager	Object Class	Serial Number
96 Bit Scheme	8	28	24	36
64 Bit Scheme	2	21	17	24

EPC Standards

EPCglobal has ratified several specifications and two standards, EPC Tag Data Standard Version 1.1 rev 1.27 and Class 1 Generation 2 UHF Air Interface Protocol Standard Version 1.0.9. These specifications and standards are user-driven because EPCglobal has subscribers from different industries that provide feedback and information to develop EPC standards. The EPCglobal Action and Working Groups help define the direction of these EPC standards. The final decision is made by the EPCglobal board of governors.

In addition to the two EPCglobal standards, the organization has established five specifications:

- 900 MHz Class 0 Radio Frequency (RF) Identification Tag Specification
- 13.56 MHz ISM Band Class 1 Radio Frequency (RF) Identification Tag Interface Specification

- 860–930 MHz Class 1 Radio Frequency (RF) Identification Tag Radio Frequency & Logical Communication Interface Specification
- Class 1 Generation 2 UHF RFID Conformance Requirements Specification v. 1.0.2
- EPCglobal Architecture Framework Version 1.0

In the future, these specifications may be turned into standards if the EPC users find it necessary. This following information gives a brief overview of the five specifications and two standards. For more information on the parameters of these standards and specifications, consult the EPC global Web site or contact the organization directly.

The first EPCglobal specification defines communications interface and protocols for 900 MHz Class 0 RFID tag operation. The information covers RF and tag requirements and operational algorithms to facilitate communications in this frequency band.

The second specification is for the 13.56 MHz Class 1 RFID tag interface specifications, which covers 13.56 MHz tag and reader functionality, including consideration of protocols and the communication interface for a 13.56 MHz tag. The specification also provides information about human exposure regulations and electromagnetic compatibility regulations.

The third specification identifies the communications interface and protocol for the 860–930 MHz Class 1 RFID tags, and covers tag and interrogator communication.

The fourth specification addresses Class 1 Generation 2 UHF RFID Conformance Specifications, including the protocols for communications and functionality associated with Class 1 Generation 2 RFID tags. This specification is essentially a list of guidelines that an 860–960 MHz RFID system must meet to be considered EPCglobal compliant. Unlike the previous specifications, the RFID conformance specifications document deals mostly with a structured format and specific compliance requirements.

The final specification that EPCglobal has established deals with architectural framework of EPCglobal, which provides a general outline for the direction the organization plans to take in developing RFID and EPC. The specification outlines the role of RFID tags, RFID interrogators, filtering, collection, EPC information services (EPCIS), and the interface and protocols associated with these roles.

International Standards

In addition to the standards developed by ISO, the Auto-ID Center, and EPCglobal, various countries have established their own standards and regulations for the use of RFID systems. The following standards are the more predominant and important standards worldwide.

USA: Federal Communications Commission (FCC)

The **Federal Communications Commission's (FCC)** Part 15 regulations, released in 2005, cover the use of radio frequency devices, which are referred to as "intentional radiators." The FCC specifications for intentional radiators are categorized by the frequency band or range on which devices operation. The regulations allow for the operation of RFID devices at the UHF range, also known as the ISM bands (Industrial, Scientific, and Medical), as specified by the FCC guidelines. For supply chain use of RFID systems, the FCC has approved the 902–928 MHz, 2400–2483.5 MHz, and 5725–5850 MHz ranges. The FCC recommends that these systems be compliant with frequency hopping spread spectrum modulation techniques, thus maximizing the power allowance for interrogator-transmitted frequencies. The complete FCC's rules can be accessed from the FCC Web site at http://www.fcc.gov/oet/info/rules.Table 7.4 identifies the sections of FCC Part 15 that pertain to RFID:

Table 7.4 FCC Part 15

Section	Topic
Part 15.1	Subpart A–Limitations on RF emissions and restricted frequency bands
Part 15.2	Subpart C–Intentional radiators
Part 15.201	Equipment authorization requirements
Part 15.202	Certified operating frequency range
Part 15.203	Antenna requirements
Part 15.204	External radio frequency power amplifiers and antenna modifications
Part 15.205	Restricted bands of operation
Part 15.207	Conducted limits
Part 15.209	Radiated emissions limits, general requirements
Part 15.213	Cable locating equipment
Part 15.217	Operation in the band 510–705 kHz
Part 15.225	Operation in the band 13.110–14.010 MHz
Part 15.240	Operation in the band 433.5–434.5 MHz
Part 15.243	Operation in the band 890–940 MHz
Part 15.245	Operation in the band 902–928 MHz
Part 15.251	Operation in the bands 2.45, 2.9–3.6 GHz

Europe: ERO, CEPT, ETSI

In Europe, the Electronic Communications Committee (EEC), European Conference of Postal and Telecommunications Administrations (CEPT), European Telecommunications Standard Institute (ETSI), and various members of the European Union have established RFID regulations. The ECC represents 48 European countries and is the most comprehensive set of regulations. The ECC deals with RFID by developing radio/communications policies and regulations for the 9 kHz to 275 GHz frequency range. ECC standards are documented and published by the European Radiocommunications Office (ERO). In 2001, the ECC also released standards for Short Range Devices (SRDs) operating at 868.0–868.6 MHz, 869.4–869.65 MHz, and 869.7–870.0 MHz.

The ETSI, created by the CEPT in 1988, developed and published telecommunications standards with the consensus of 55 member countries. It has approved the use of the UHF range, but with limitations on power, bandwidth, and duty cycle. These regulations specify a maximum of 500 mW of transmitted power in a band of 200 kHz; however, ETIS is currently considering an amendment that would allow a maximum of 2 watts of transmission power.

China: SAC

The Standardization Administration of China (SAC), collaborating with China's Ministry of Information Technology, has developed China's RFID regulations. In 2004, the SAC established the National RFID Tag Standard Working Group, which outlined the *China*

National RFID Tags Standards. These standards are a compilation of ISO standards and standards developed by the SAC including:

- ISO 15693
- Physical specifications for identification cards, non-contact IC cards, and protocol concerns
- Input/output port and initialization standards for identification Cards, non-contact IC cards, and protocol cards
- Anti-interference and transporting standards for identification cards, non-contact IC cards, and protocol cards
- Application standards for electronic labeling and terminal reading

Table 7.5 lists the common frequency ranges and the countries that permit operation at specific frequencies.

Table 7.5 Internationally Used Frequencies

Frequency	Frequency Regions/Country
125-134 kHz	United States, Canada, Japan, and Europe
13.56 MHz	United States, Canada, Japan, and Europe
433.05-434.79 MHz	United States, most of Europe, and under consideration in Japan

865–868 MHz	Europe
866–869 MHz	South Korea
923–925 MHz	South Korea
902–928 MHz	United States
952–954 MHz	Japan (for passive tags starting in 2005)
2400–2500 GHz	United States, Canada, Europe, and Japan
5.725–5.875 GHz	United States, Canada, Europe, and Japan

Standardization and Security

Security is a major concern, especially for technology as pervasive as radio frequency. EPC standards and country requirements have mandated security measures for the encryption of RFID data. Encoding and decoding are serious issues with RFID systems, and since the medium for transmission is air, technicians and people working on these systems must be able to make the data secure.

Modern **authentication protocols** work by checking a **cryptographic key**. This key is vulnerable to thievery, so suitable algorithms are often used to prevent the key from being uncovered or cracked. In order to establish a secure RFID system, the following infringements must be considered:

- Unauthorized access or reading of data carriers in an attempt to duplicate or modify data

- Insertion of illegitimate foreign data carriers within the interrogation zone of a reader with intentions of a malicious nature. This is typically done to gain access to a building or receive services without payment

- Eavesdropping into radio communications in order to emulate genuine data carriers

Cryptological functions are expensive, but provide effective enhancement for high security applications. Applications that do not require a security function, such as industrial automation, would be made unnecessarily expensive with the addition of cryptological functions. For this reason, only high-security applications such as ticketing and payment systems typically use cryptological functions.

There are several methods used for authentication between interrogator and tag. Mutual Symmetrical Authentication and authentication using derived keys are the two most popular for RFID systems.

Mutual authentication between interrogator and tag is based on the three-pass authentication principle, where the participants in the communication check the other party's knowledge of a cryptological key (Figure 7.3). The process requires all tags and receivers to have the same cryptological key. In order to protect the system from false manipulation, a tag is initially assumed by the interrogator to be a foreign tag until it has been authenticated.

The interrogator begins the authentication procedure by sending a GET_CHALLENGE command to the tag. A random number R_A is then generated in the tag and sent back to the interrogator. Next, the interrogator generates a random number R_B. Using the cryptological key K and a common key algorithm, the interrogator calculates an encrypted data block (Token 1), which contains both random numbers and additional control data, and sends the block back to the tag. Token 1 is decrypted within the tag and the plain text random number R_A is compared to the previous R_A. If the figures match, the tag generates another random number R_{A2}, which is used to calculate a second encrypted data block (Token 2). Token 2 contains R_B and is sent back to the interrogator along with control data. The interrogator decrypts Token 2 and checks if R_{B2} matches with the previous R_B. If the two match, the authentication is complete.

Mutual authentication is ideal for applications with a minimal numbers of tags, but it is not feasible for applications with vast numbers of tags because the probability that the key will be discovered increases with larger quantities of tags. This problem can be significantly improved by giving each tag a different cryptological key.

In the derived key method, the serial number of each tag is read out during production. The tag is initialized by calculating a key K_X using a cryptological algorithm and a master key K_M. As a result, each tag receives a key linked to its own ID number and the master key.

The actual authentication process starts with the interrogator requesting the ID number of the tag. In the security authentication module (SAM) in the interrogator, the tag's key is calculated using the master

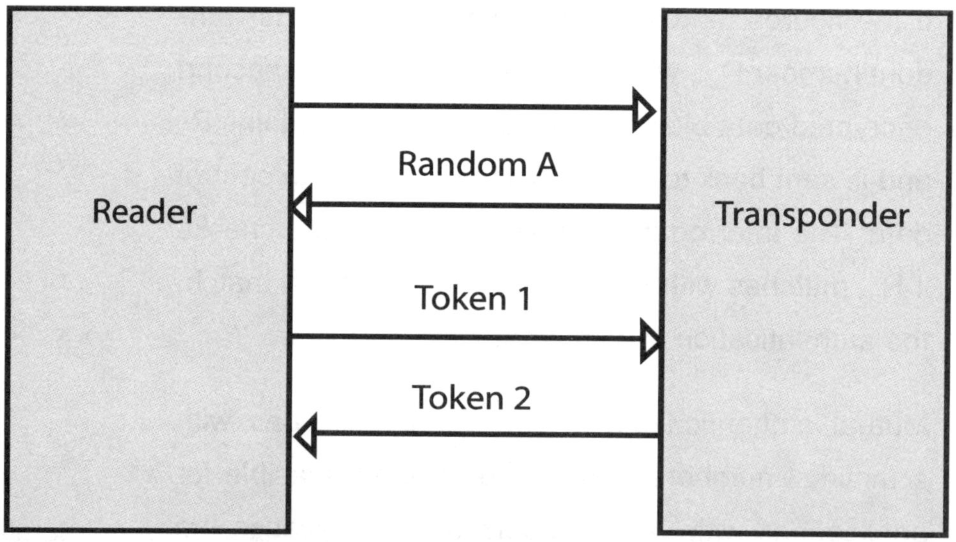

Figure 7.3 Mutual Authentication

key. This key is then used for authentication. This SAM normally incorporates a cryptoprocessor, which means that the stored master key can never be read.

Summary

The standards and regulations of ISO, EPCglobal, and the FCC, as well as the guidelines of regional bodies and governments, limit and define the selection, deployment, and operation of RFID systems. When implementing a system, a technologist must be conscious of the specific rules and regulation that will apply to system's frequency range, components, parameters, protocols, etc. In each case the most current guidelines from the regulating body should be referenced. The following Web sites are a helpful starting point:

Aim Global	http://www.aimglobal.org/rfid
EPC Global	http://www.epcglobal.com
FCC	http://www.fcc.gov
ISO	http://www.iso.org
ERO and ECC	http://www.ero.dk/ecc
CEPT	http://www.cept.org/
ETSI	http://www.etsi.org

Review

Questions

1. The ISO 18000 Series standards approve the following air interface protocols *except*?

   a. 13.56 MHz

   b. 2. 45 GHz

   c. 860–960 MHz

   d. 5.8 GHz

2. Which ISO standard would a technologist consult if he or she is configuring a system using contact-less integrated circuit cards?

   a. ISO 14443

   b. ISO 10536

   c. ISO 11784

   d. ISO 10373

3. ISO 10536 covers the characteristics, dimensions, and transmission protocols of?

   a. interrogators

   b. close-coupled cards

   c. proximity cards

   d. Chipless transponders

4. The United States, Canada, Europe, and Japan have all accepted ________ as a legal operation range.
   a. 5.725–5.875 GHz
   b. 902–928 MHz
   c. 865–868 MHz
   d. 866–896 MHz

5. Which of the following is an air interface protocol?
   a. ISO 18000-5
   b. ISO 10536
   c. ISO 14443
   d. ISO 19763-3

Discussion

1. An ultrahigh frequency system originally deployed in France is being moved to Texas. How might the standards for the system change?

2. Explain the specific regulations and standards that apply to RFID systems used with animal identification.

3. What specific standards apply to human exposure regulations? Explain how to keep a system compliant with these regulations.

Chapter 8
Applications

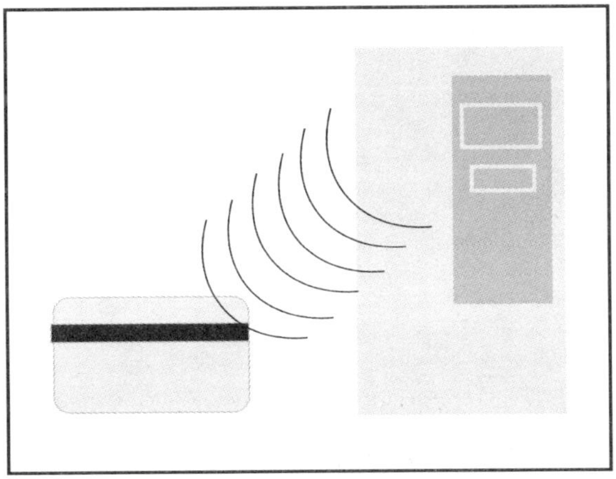

Chapter 8: Applications

The following case studies apply RFID principles to real-world system applications. There is an infinite number of RFID applications, but the applications selected for this chapter are applied RFID+ concepts.

Part 1
Smart Cards

In Europe and Asia, one of the most popular RFID applications is Smart Cards. The actual construction of the cards was described in Chapter 5: RFID Tags. To the user, these tags look and feel like any other credit card. But the RFID IC and antenna, which are encased in the plastic, are read much faster than the magnetic strip on a traditional credit card, and a user does not need to remove the card from his or her purse or wallet for it to be scanned.

ISO 14443 standard, Proximity Coupling Smart Cards, specifies a 13.56 MHz operating frequency. Amplitude Shift Keying is the modulation used to signal from the interrogator to the card; the response from the card uses on/off keying for ISO 14443 type A and binary phase shift keying for ISO 14443 type B. Both types have a data transfer rate of 106K bits/sec. Typical interrogation zones extend 10 cm from the antenna. This read range is smaller than UHF systems, but more secure, making Smart Cards an important consideration for financial transactions.

A popular application for Smart Cards is public transportation. Some systems work like a debit card, where the card is "charged" and only the amount left on the card and an account number is stored on the card. Others store a history of the recent trips charged to the account. All systems require an interface between the interrogators at points of entry and departure from the transit system and a central computer network that keeps track of the fares.

The primary benefit to Smart Card users is the speed at which they can negotiate the transit system. Rather than feeding a magnetic stripe card through a reader and pulling it out again when a gate opens, a user can leave a card in their purse, for instance, and simply pass the purse over the antenna. The data transfer is so fast that the user does not need to pause for the gate—they just walk or drive right through. The Smart Card can be up to five times faster than a conventional magnetic stripe card (Figure 8.1).

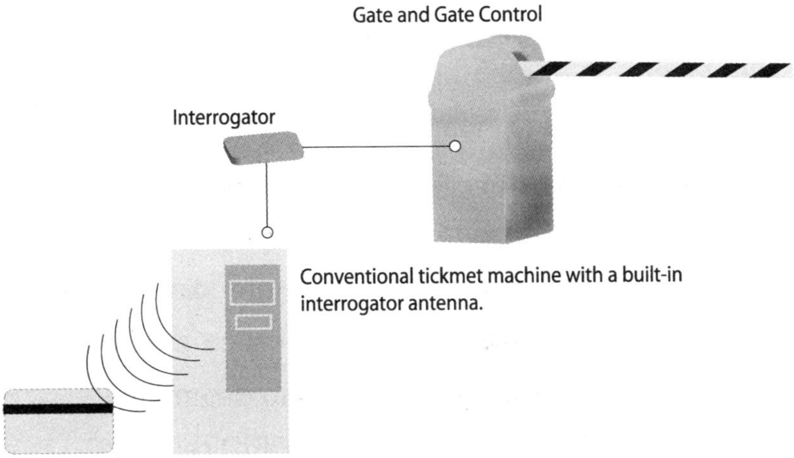

Figure 8.1 Smart Card in Public Transportation

An RFID system also has benefits to the transit authority. RFID interrogators have no moving parts, unlike the magnetic stripe readers they replace. This means fewer maintenance expenses. Additionally, unlike paper cards that have a lifespan of only a few weeks, Smart Cards are designed to last about ten years.

Currently, the market for Smart Cards in public transit is for daily commuters, who are willing to pay a small fee to purchase a Smart Card and use it for years. Until the costs of disposable, one-use card types decrease, most transit companies will continue using magnetic stripe cards concurrent with Smart Card technology.

Part 2
Printers for RFID labels

An RFID label printer can encode RFID tags, print barcodes, and/or create human readable text all in one process. An RFID label is an RFID tag and printable label (such as a barcode label applied to a case of product) combined into one. There are distinct advantages to RFID labels (sometimes called smart labels) over plain RFID tags: a human can determine the contents of a labeled box without an RFID reader, and the label can work with applications that use the older barcode technology.

An RFID label (Figure 8.2) is a sticker with an RFID tag inlay sandwiched into it. It is composed of several layers:

- Label top coat (printable surface)
- Label material (paper-like)
- Tag inlay (consists of the tag IC and tag antenna)
- Adhesive (must be compatible with package material so it will stick)
- Carrier sheet with liner release coating

RFID labels have all the concerns and requirements of all RFID tags: class types, frequency range, RF readability, etc. They also have all the quality considerations of optical labels, such as printability, adhesion, robustness, and longevity. These concerns come together both in label selection and label placement.

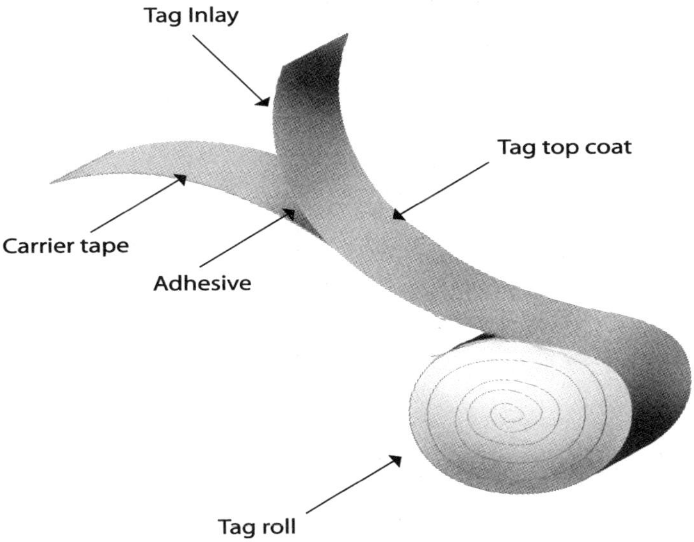

Figure 8.2 Label Diagram

Label Selection

When selecting an RFID label, it is important to consider following:

- Will the lablel stick to the package (correct adhesive)?

- Will the label withstand the environmental factors (correct label material)?
- Can the label be printed by the appropriate method (thermal is most common) for your printer?
- Is the the label the correct size for the package and the amount of information to be printed on it?
- Will the tag inlay operate at the appropriate frequency?
- Is the tag inlay of the correct class and specifications to work with the interrogators that will be reading the data?

Label Placement

Placement of the tag on a package involves the considerations of RFID tags (described in an earlier chapter) and readability by humans and machines (such as barcode readers). Whereas RFID tags may be imbedded inside an item, visually readable labels must be within line-of-sight of the reader. In many instances, placing an RFID tag where it will present itself parallel to an antenna will also present the label toward a barcode reader.

Encoding

The process of encoding a tag inlay inside an RFID label is the same as encoding other RFID tags. Most interrogators used for reading tags may also be configured to program. In order to make sure that every tag is unique, an interrogator can only program one tag at a time. To make sure that only one tag is addressed during a write cycle, it is important only one tag is in the interrogation zone.

More power and time is needed to write to a tag than to read from it. Programming tag inlays inside a printer offers advantages over a more open space system. Since the labels feed through the printer in an orderly manner, the antenna can be set up and tuned so that the interrogation zone only excites one tag at a time. The proximity of tag to antenna allows inductive coupling between the two antennas, thus increasing the power transfer and efficiency. The printer can also indicate on the label which tags could not be programmed and verified, thus making it obvious which tags should be replaced.

Printers can be set up to apply labels as part of an assembly or packaging line, or used in a manual "slap and ship" process. In an automated process, the printer will interface with the inventory management software to acquire the data to encode on each inlay and report back what serial numbers have been assigned. In a manual process, the operator might use a barcode reader to acquire data from an existing barcode label. This information is communicated to the printer (usually through a computer interface with the barcode reader) and then programmed into the RFID label. The printer serializes the tag and prints the encoded information in barcode and/or human readable form onto the label. Finally, the operator places the label on the item.

Part 3
Improving Agribusiness: Supply Chain Efficiency by Utilizing RFID

Case Study Courtesy of **BlueBean, LLC,**
www.BlueBeanRFID.com

During harvest time, a major agribusiness company that produces pistachio nut packages receives up to 21 million pounds of nuts per day. The nuts are shipped in bulk in trailers. To determine the amount of nuts received per truckload, the full trailer is weighed and the tare (empty weight of the trailer) is subtracted to determine the "net green weight" of the nuts.

In this application, each of the 2000 trailers was equipped with an RFID tag containing a unique identification number (Figure 8.3). The system operates at 915 MHz. Because the metal of the trailer could interfere with the tag operation, each tag was applied with a 1/8-inch "spacer." Durable tags were required, since the trailers remain outdoors in desert conditions year-round.

Two interrogators are installed at each scale to read the tags. The trailer's unique identifier number is used to retrieve the trailer's tare weight and other information from a database. The tare weight is used to calculate the net weight of the nuts; the other information is used for inventory management.

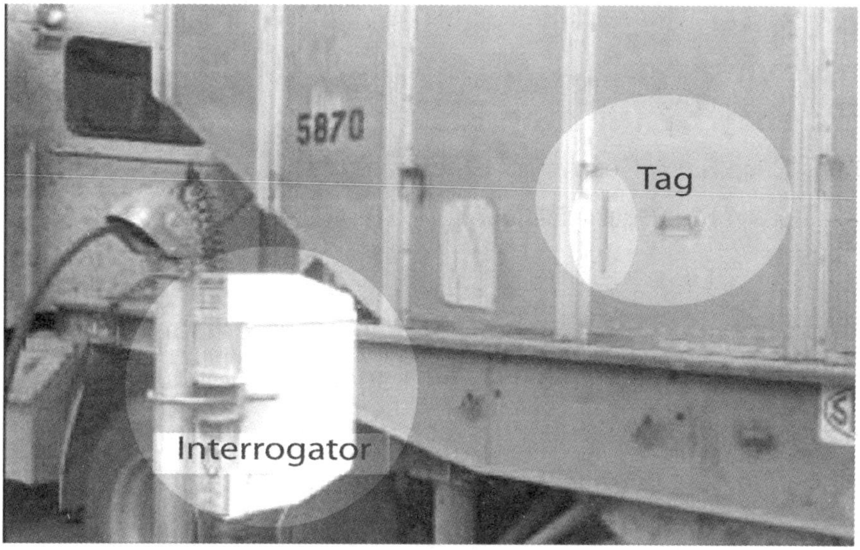

Figure 8.3 Trailer Tag and Interrogator

Use of RFID has the following advantages over human readable or barcode solutions:

- Increased load processing efficiency by 60%
- Improved equipment utilization by 30%
- Reduced harvest headcount by 30%
- Reduced human error
- Reduced manual entry
- No concern of visual identification wearing off or being obscured

- Improved information management

The increased efficiency of data collection has benefits in the rest of the processing operation. Furthermore, use of the RFID system allows asset tracking of the trailers themselves.

Part 4
Furniture (Asset) Tracking

Case Study Courtesy of Northern Apex, www.northernapex-rfid.com

RFID can be a viable method of tracking expensive assets. As the value of an individual asset increases, return on the investment for an RFID system also increases.

A Veterans' Administration hospital gained cost savings by tracking the hospital's furniture. Although office furniture is not critical to the hospital's patient care mission, it is still necessary to keep inventory secure and to be able to locate it. Furniture, especially chairs on wheels, may easily be moved to other offices or locations without updating the database being updated accordingly.

A 13.56–MHz passive tag was attached under each piece of furniture. The tag is read-write, and stores two kilobytes of data. For this application, information stored in the tag's memory includes

- Asset information (ID, description, manufacturer, etc.)

- Warranty information (in-service date and warranty expiration date)
- User characteristics (such as the assigned location)
- Maintenance history (modifications or repairs)

A PDA equipped with a portable RFID interrogator and antenna is used to read the tags and temporarily store information (Figure 8.4). Software on the PDA allows it to interface with the attached interrogator. The antenna must pass within a few inches of the passive tag to read it, so a human must go into the room to scan the furniture. This should help with privacy concerns sometimes associated with RFID installations in public places.

Figure 8.4 Interrogator PDA

Part 5
Receiving Conveyor System

Case Study Courtesy of BlueBean, LLC,

www.bluebeanllc.com

A common application inside a warehouse is a conveyor to move and sort packages. Using RFID to identify packages is an obvious solution. Here, a RF cage contains four antennae all controlled by the same interrogator (Figure 8.5). The antennae are aimed to maximize the zone inside the RF cage. By placing circularly polarized antennae on top, sides, and bottom of the cage, tags on any side of box can be read. Tags that present themselves on the front or back of the box are read as the box is entering or leaving the cage.

Figure 8.5 Four Antenna Configuration

The rollers inside the cage are made out of plastic rather than metal to eliminate any scattering effect from reflections from the rollers. This also allows the antenna under the conveyor to "see" the tags on boxes riding down the conveyor.

Wire mesh is used around the interrogation zone to hold the RF signals inside the zone. The mesh prevents this RFID system from interfering with nearby RF systems and those systems from interfering with the conveyor system. Without it, the interrogation zone may extend down the conveyor, and the system will have to contend with more than one box at a time. Although the interrogator can handle the tag information, the system must recognize the order of the packages so that they can be sorted properly. By reading only the one box inside the station, the system can accurately determine each box's destination.

The interrogator interfaces with a PLC that controls the conveyor on the entry side, the conveyor inside the cage, and the conveyor on the exit side. It also controls a sorting arm, indicator lights, and some networking functions. Signals from sensors on the entry and exit conveyors tell the PLC where packages are so the PLC can determine whether or not to run each section of the conveyor (Figure 8.6).

Figure 8.6 Enclosure with PLC and I/O Connections

In this application, the sorting arm (Figure 8.7) is used to push some packages onto a separate conveyor, or let the boxes continue on down the same line. In order to make this determination, the tag information from the interrogator must be passed to middleware that requires a customer data base. Information from the data base is then used by the PLC to determine which path the box should follow. The entire process to move a box into the zone, read the tag, query the data base, make a decision, and move the box to the proper conveyor must be done without any pause in the line.

Figure 8.7 Sorting Arm

APPENDIX

Appendix A
Practice Exam Questions and Answers

The following are the multiple choice, short answer, and true and false questions and answers for chapters 1 through 7.

Chapter 1: Primer

1. What units are used to measure inductance?
   a. Ohms
   b. Volt
   c. Watt
   d. Henries
2. Which will have a longer wavelength?
   a. signal of 10 MHz
   b. signal of 50 MHz
   c. signal of 100 MHz
   d. signal of 200 MHz
3. The correct relationship between period and frequency is?
   a. f = 1/T
   b. T = 1/F
   c. fT = 1
   d. All of the above
4. In an inductor, what is the cause of opposition to the circuit?
   a. Reluctance
   b. Resistance
   c. Capacitance
   d. None of the above
5. What VSWR will result if a 50-ohm load is connected to an antenna with a 10-ohm feed point impedance?
   a. 1
   b. 1.5
   c. 5
   d. .5
6. What is the resonant frequency of a series circuit made with a 10-mH inductor and a 3-uF capacitor?

   919 Hz
7. An RF signal bounces off a metal frame. This is an example of _____.
   a. refraction
   b. diffraction
   c. reflection
   d. sky waves
8. An RF signal bends around a wall. This is an example of _____.
   a. refraction
   b. diffraction
   c. reflection
   d. ground waves
9. When RF energy is absorbed by an object, it is usually transformed to _____.
   a. light energy
   b. induction
   c. magnetic field
   d. heat energy
10. Which has a higher attenuation constant: insulators or conductors?

    conductors
11. An RF signal travels faster through ______.
    a. air
    b. water
    c. glass
    d. rubber
12. True or False: All the power in an RF signal is contained in the carrier.

    false
13. True or False: Only AM modulation creates side bands that carries information.

    false
14. Spread spectrum systems, when compared to narrowband systems _____.
    a. use more bandwidth
    b. use less power
    c. are less prone to interference
    d. all of the above
    e. both a & c

15. Amplitude Shift Keying is an example of _____.
    a. amplitude modulation
    b. digital modulation
    c. spread spectrum modulation
    d. both a and b

Chapter 2: System Lifecycle

1. Which of the following will a blueline drawing not indicate?
    a. Antenna radiation patterns
    b. Doors and windows
    c. Existing RF system installations
    d. Power and network outlets
2. A site analysis will determine _____.
    a. the best placement of RFID interrogator antennas
    b. potential sources of RF interference
    c. readability of a specific tag in the system where it will be used
    d. where to place tags on an item to be tracked
3. A spectrum analyzer shows _____.
    a. voltage in dB versus time
    b. power in watts versus frequency
    c. voltage or power in dB versus frequency
    d. power in watts versus time
4. A "Full Faraday Cycle" analysis is to be performed to prepare for an RFID installation that will use tags that operate at 13.56 MHz. The best frequency span setting for the spectrum analyzer is ______.
    a. 10 MHz, centered at 10 MHz
    b. 3 MHz, centered at 13.5 MHz
    c. 13.56 MHz, centered at 6.75 MHz
    d. 14 MHz, centered at 13.56 MHz
5. Cables are installed inside conduit because _____. (select all that apply)
    a. more cables can be held in a smaller space using conduit than other methods.
    b. conduit shields the cables from RF interference.
    c. cables can pose a tripping hazard.
    d. the National Electric Code requires it.
6. In RFID systems using passive tags, the operation starts with a "wakeup" signal generated by the _____.
    a. tag antenna
    b. interrogator antenna
    c. tag
    d. interrogator
7. Select all the potential sources of ESD from the list below:
    a. Antennas
    b. Boxes moving down conveyors
    c. Grounding cables
    d. Label dispensing
    e. Paper handling
    f. Wireless LAN hubs
8. What component is most sensitive to ESD?
    a. Tag
    b. Interrogator antenna
    c. Interrogator
    d. Network hub
9. List at least three sources of possible RF interference that would be identified on a blue-line drawing as preparation for an RFID installation design.
10. True or False: One shift of operation will be adequate time for a Full Faraday Cycle analysis.
    false; some devices may be sensitive to higer power RF signals like those close ot the interrogator antenna.
11. Where should antenna be mounted to collect data for a site analysis? (select all that apply)
    a. In the middle of the proposed interrogation zone
    b. On all four corners of the proposed interrogation zone
    c. Random places around the site
    d. Corners of the facility
12. Describe how an antenna and a spectrum analyzer may be used to locate a source of RF interference.
13. Which of the following would be the preferred method of mounting an antenna so as not to interfere with its radiation pattern?
    a. Plastic tie-wraps
    b. Plastic screws
    c. Metal bolts
    d. Metal brackets
14. What current draw would be expected for an RFID interrogator and associated hardware?
    a. 100 mA
    b. 1 A
    c. 3 A
    d. 100 A

Chapter 3: Frequency Ranges

1. RFID "chips" implanted in animals and pets operate in which frequency range?
   a. Low
   b. High
   c. Ultrahigh
   d. Microwave
2. Tags used on the ears of cattle to interface with interrogators about 1 meter away as the cattle walks through portals operate in which frequency range?
   a. Low frequency
   b. High frequency
   c. Ultrahigh frequency
   d. Microwave frequency
3. Low frequency system read ranges are limited to a half meter because _____.
   a. of the RF physics of low frequencies
   b. of the regulations that limit power output of low frequency systems
   c. if frequencies exceed ½ meter they are high frequencies
   d. RFID components cannot transmit at any other frequency
4. Which frequency range operates the best around metals?
   a. Low
   b. High
   c. Ultrahigh
   d. Microwave
5. Which frequency range has the fastest data transfer rate?
   a. Low
   b. High
   c. Ultrahigh
6. Which RFID system frequency is internationally accepted?
   a. 134.2 kHz
   b. ***3.56 MHz***
   c. 915 MHz
   d. 2.5 GHz
7. A library wishes to implement an RFID system. For security and privacy issues, it is desirable that the read range of the tags be approximately 1 meter. The cost per tag should be minimal. Which frequency range should be used?
   a. low
   b. high
   c. ultrahigh
   d. microwave
8. The frequency range that has a read range of 3 meters is the ______.
   a. low frequency range
   b. high frequency range
   c. ultrahigh frequency range
   d. microwave frequency range
9. The 125-134 kHz frequency region is allowed in which of the following areas?
   a. the United States
   b. Canada
   c. Japan
   d. Europe
10. The _____ frequency range is accepted worldwide and operates well around water and metals.
   a. low
   b. high
   c. ultrahigh
   d. microwave
11. Systems on the _____ frequency range are generally used for toll collection and logistics applications.
   a. low
   b. high
   c. ultrahigh
   d. microwave

Chapter 4: Antennas

1. Which of the following devices can be used to visually display wave signals?
   a. Transmitter
   b. Oscilloscope
   c. Stethoscope
   d. GUI
2. Antenna Gain is a measurement of _____.
   a. the total intensity of an antenna's signal
   b. a ratio of the power an antenna receives and transmits

c. the ability of an antenna to focus radio waves in a particular direction

d. the average bandwidth of the waves an antenna propagates

3. Directivity is also known as ______.

a. ERP

b. field direction

c. maximum directive gain

d. directive power

4. The main lobe of directional antenna is the _____.

a. maximum VSWR point of radiation

b. direction of least radiation from the antenna

c. direction of maximum radiated field intensity from the antenna

5. An antenna with low radiation resistance will have a high or low efficiency?

high

6. True or False: Antenna efficiency is always less than 1 (or 100%).

true

7. True or False: Directive gain is always less than 1.

false

8. True or False: Power gain is always less than 1.

false

9. Calculate the power gain in decibels for an antenna with directive gain of 5 and efficiency of .5 (50%).

10log(2.5) = 4dB

10. A power specified in dBm means that _____.

a. the actual power was multiplied by 10-3 before taking the logarithm

b. the actual power was expressed in mW before taking the logarithm

c. there is no difference between dB and dBm, so it has no significance

d. the directive gain in dB was added to the result

11. The EIRP is a measure of ______.
(select all that apply)

a. the directivity of an antenna

b. the efficiency of an antenna

c. power density at a particular point in space

d. power radiated by a specific antenna

12. Calculate the power density 1 m away from an antenna with EIRP of 3 dBm.

convert EIFP to mW: P(mW):
P(mW) = antilog (3dBM/10) = $10^{.03}$ = 2mW

use the equation for power density (4-6c).

13. A receiving antenna has a power gain of 0.5, and captures 30 mW of power at 900 MHz. Calculate the actual power density of the 900 MHz signal.

3E9m/sec/900E6Hz = 3.33m
use equations 9-9 and 9-10a
Ac = .5$(3.33m)^2$/(4pi) = 44m^2
use equation 9-10c
Rearrange Eq 9-9 Pdensit = Pcap/Ac = 30mW/ .44m^2 = 68mW/m^2

14. How much current is needed to create 100 mV across a 50 Ω antenna? How much power is this?

I = 100mV/50Q = 2mA
P= V^2/R = $(100mV)^2$/50 =

(.1.1)/50 = .0002W = 200μW

Chapter 5: Tags

1. A tag's antenna typically needs to measure _______ of the original frequency's wavelength.

a. half

b. 2/3

c. 25 %

d. 0%

2. RO tags provide a ________ level of security as RW tags.

a. lower

b. higher

c. similar

3. Which of the following is most RF-absorbent?

a. Bubble wrap

b. Plastic bottles of water

c. Empty pop cans

d. Metal doors

4. Which of the following tag types use internal battery power (circle all that apply)?

a. Semi-passive

b. Active

c. Passive

d. Semi-active

5. What will the ¼ wave antenna length be for an RFID tag that operates at 134 kHz?

6. Match the tag type with the correct class:

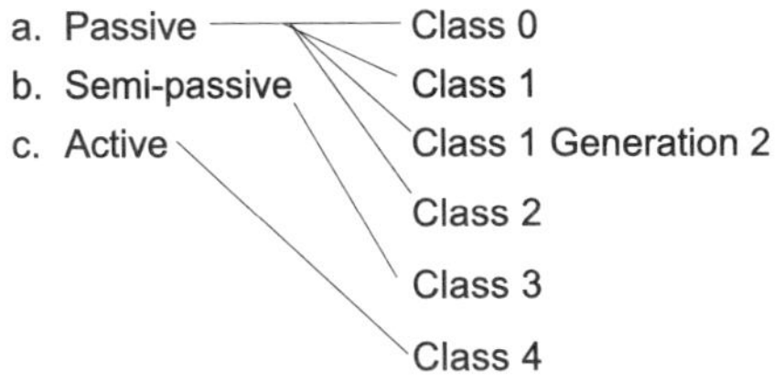

7. What class of tags includes integrated sensors?
   a. Class 0/1
   b. Class 1 Generation 2
   c. Class 2
   d. Class 3

8. What class of tag would be used to collect information from other tags in the vicinity?
   a. Class 0/1
   b. Class 1 Generation 2
   c. Class 2
   d. Class 3
   e. Class 4

9. Select the best tag placement on a case of bottled water from the following choices: (see figure in file "tags on cases" in chapter 5, pg 173)

   Placement A

10. Select the best tag placement on a book that will be read both on the shelves and as it is "checked out."
    a. In the middle of inside front cover
    b. On the top left side of the inside front cover
    c. In the middle of inside back cover
    d. On the spine

11. Which of the following is the least RF absorptive?
    a. Paper
    b. Glass
    c. Human body
    d. Drinking water

Chapter 6: Interrogators

1. Which of the following should not be used for the connection between the interrogator and host computer?
   a. Ethernet
   b. IEEE 802 wireless
   c. ISO 18000
   d. RS232 null modem

2. What primary type of interrogator would be utilized at an incoming dock door?
   a. Handheld
   b. Fixed position (in portal installation)
   c. Mobile mount
   d. Interrogator/encoder

3. What secondary type of interrogator would be utilized at a dock door?
   a. Handheld (for back up verification)
   b. Fixed position
   c. Mobile mount
   d. Interrogator/encoder

4. An interrogator uses mono-static antennas. This means that ______.
   a. it can only operate at one frequency and it does not support frequency hopping
   b. it can only operate in full-duplex mode
   c. it can only operate in half-duplex mode
   d. it can operate in full- or half-duplex mode

5. True of False: Firmware should typically be edited by the technologist to adapt the interrogator for each installation.

   False

6. Which of the following is typically not done by edgeware?
   a. Consolidate multiple reads from one tag
   b. Configures power level of the interrogator
   c. Configures polling method
   d. Interfaces with inventory management software

7. Name three considerations when installing multiple interrogators in close proximity to one another.

   1. Proper tuning of each interrogation zone to eliminate interference.
   2. RF interference reducing read rate/RF interference totally blocking tag

3. Dense interrogator mode option for Gen 2

4. Network traffic

8. A null modem cable _____.
   a. is the same as a coaxial cable
   b. crosses the transmit and receive lines
   c. is always a fiber-optic cable
   d. is not used with interrogator components

9. When an interrogator(s) is connected to a network protocol, it works with the _______ software to communicate information to the network.
   a. firmware
   b. middleware
   c. edgeware
   d. host computer

10. Which of the following specifications can usually be found on a vendor's interrogator data sheet?
    a. Configuration software availability
    b. Available upgrades
    c. Supported tag protocols
    d. All of the above

11. The ______ is affixed to a vehicle and scan pallet tags.
    a. mobile-mount interrogator
    b. handheld interrogator
    c. dense interrogator
    d. fixed-position interrogator

12. Middleware is responsible for _______.
    a. the configuration of an interrogator's function the operation of a interrogator's microprocessor
    b. the graphic presentation of data on the interrogaior's GUI
    c. the collection, management, and flow of data

Chapter 7: Regulations and Standards

1. The ISO 18000 Series standards approve the following air interface protocols *except*?
   a. 13.56 MHz
   b. 2. 45 GHz
   c. 860–960 MHz
   d. 5.8 GHz

2. Which ISO standard would a technologist consult if he or she is configuring a system using contact-less integrated circuit cards?
   a. ISO 14443
   b. ISO 10536
   c. ISO 11784
   d. ISO 10373

3. ISO 10536 covers the characteristics, dimensions, and transmission protocols of?
   a. Interrogators
   b. Close-coupled cards
   c. Proximity cards
   d. Chipless transponders

4. The United States, Canada, Europe, and Japan have all accepted ________ as a legal operation range.
   a. 5.725–5.875 GHz
   b. 902–928 MHz
   c. 865–868 MHz
   d. 866–896 MHz

5. Which of the following is an air interface protocol?
   a. ISO 18000-5
   b. ISO 10536
   c. ISO 14443
   d. ISO 19763-3

Appendix B
Abbreviations

ASK	Amplitude shift keying
CEPT	European Conference of Postal and Telecommunication Administration
DSB SC	Double Sideband Full Carrier
DSB SC	Double Sideband Suppressed Carrier
DPSK	Differential phase shift keying
EPR	Effective radiated power
ETSI	European Telecommunications Institute
EU	European Union
FCC	Federal Communications Commission
FHSS	Frequency-hopping spread spectrum
FM	Frequency modulation
FSK	Frequency shift keying
HF	High frequency
IC	Integrated circuit
ISM	Industrial, Scientific, Medical
LC	Inductive-capacitive or Inductor-capacitor
LF	Low frequency
Q	Quality factor
QPSK	Quadrature phase shift keying
RFID	Radio Frequency Identification
SSB-SC	Single-Sideband Suppressed Carrier
UHF	Ultrahigh frequency
UWB	Ultra wide band

Appendix C
Glossary

absorption

When a material retains energy from an antenna, resulting in a loss of signal quality.

active tag

A tag that has a transmitter to send information back to an interrogator instead of reflecting back the interrogator's signal. Active tags use a battery to power their transmission.

amplification

The augmentation of an antenna's signal strength.

amplitude

The highest or greatest value of a periodic wave. Amplitude is measured along a wave vertical axis.

amplitude modulation

Modifying the amplitude of a radio wave.

amplitude shift keying (ASK)

A modulation process that uses two or more discrete amplitude levels or heights to represent binary values 1 and 0.

antenna

Device that radiates and receives radio signals. In RFID systems, both transponders and interrogators use antennas to transmit and receive radio signals.

antenna gain

A measure of an antenna's ability to focus radio waves in a particular direction, which is a ratio of power transmitted to power received expressed in dB.

Antenna Input Impedance

Determines the ability of an antenna to accept power for a amplifier or other source.

application layer software

Layers of software that are used to communicate, sort, translate, and interpret data.

attenna radiation pattern

The measurement of an antenna's signal that is used to determine the outline of the interrogation zone.

attenuattion coefficient

A factor of a material's capacity to absorb or reduce power.

backfire antenna

A directional antenna ideal for outdoor use because of its compact and durable design.

backscatter

Radiation scattering in directions reverse to the incident beam.

bandwidth

Antenna frequency spans from the highest to lowest assigned frequency band.

barcode

A unique 14-digit code used to identify objects. Barcodes are often labels that can be scanned by a computer.

beamwidth

The angular separation between the two half-power points on an antenna's major lobe. Bandwidth measures antenna focus or concentration.

bi-static antenna

An interrogator antenna configuration that use two antennas. One of the antennas transmits signals while the other receives signals.

blade antenna

Knife- or blade-shaped antennas that are usually housed in plastic casing. Blade antennas are used to augment signals or pick up low signals.

characteristic impedance

The value of impedance of a transmission line that is lossless. Character impedance does not account for signal depreciation which exists in all real-world environments.

contact-less Smart Card

A credit-card-like plastic tag that contains a integrated chip.

cryptological key

A security algorithm used to protect systems. Most cryptological keys use authentication protocols that require authentication of both tags and interrogators.

coupling

The linking or connection of magnetic fields. Coupling allow circuits to radiate and transfer energy.

decibel (dB)

A measure of the gain of an antenna.

dense interrogator mode

When two or more interrogators operation within close proximity, dense interrogator mode is used to avoid interference between the interrogators.

diffraction

The bending of waves around objects.

directional antenna

An antenna that radiates signal is one direction.

directive gain (directivity)

The property of transmitting or receiving energy more strongly in some directions than in others.

disk/coin tags

Tags, usually passive, that are enclosed in a round disk or coin shaped casing. Disk/coin tags are one of the most common tag formats

dome antenna

A multi-directional antenna that is used in situations where a wide interrogation zone is required.

duplex

A channel capable of transmitting data in both directions at the same time. (Half duplex is a channel capable of transmitting data in both directions, but not simultaneously.)

earth ground

A facility conductor that safely grounds the AC power line.

edgeware

Host computer software that is used to configure an interrogator's read rate, power level, and frequency. Edgeware also functions as a data conduit between the data collection application and firmware.

Effective Isotropic Radiated Power (EIRP or ERP)

A measure of antenna signal strength, EIRP is the mathematical product of the power supplied to an antenna and the antenna's gain. EIRP is expressed as dBW.

efficiency

The ratio of the output to the input of any system

Electronic Product Code (EPC)

A serial, created by the Auto-ID Center, that will complement barcodes. The EPC has digits to identify the manufacturer, product category and the individual item.

Electrostatic discharge (ESD)

Any source of electrical dissipation that can be harmful to equipment. ESD can occur any place where material rub or separate consistently.

far field

The area extending from an antenna wherein the electric and magnetic fields are in phase together.

Federal Communications Commission (FCC)

U.S. governmental body that regulates use of the radio frequency spectrum.

Faraday's Law

Statement that accounts for the induction of voltage consequent a shift of flux in a circuit's magnetic fields.

field intensity

The vector sum of all the forces exerted by an electrical or magnetic field (on a unit mass or unit charge or unit magnetic pole) at a given point in the field

firmware

Software that operates in an interrogator, using the interrogator's Read-Only-Memory.

fixed-position interrogator

A stationary interrogator often used for conveyor and portal configurations.

flux line

The lines of direction in a magnetic field. Faraday observed that shift in flux lines in the magnetic field induces voltage.

frequency

The number of repetitions of a complete wave within one second.

frequency hopping

A modulation technique that rapidly switches or hops sequentially between the frequency channels.

frequency modulation

The alteration of a high-frequency carrier in relation to a baseband signal. There are two primarily types of frequency modulation: amplitude modulation: AM and frequency modulation: FM.

frequency shift keying

The variation of a signal between two frequencies according to the binary inputs of the system.

frequency spectrum

The distribution of signal amplitudes as a function of frequency.

full duplex

One of two interrogator modes. In full duplex mode, an interrogator and send and receive information at the same time.

Full Faraday Cycle Analysis

An analysis lasting a full business cycle (24 or 48 hours) that identifies possible source of RF interference at an RFID system site.

gain

The ratio of a system's signal output to signal input.

ground loop

The unfavorable occurrence of current traveling along ground wires in a circular path.

half duplex

One of two interrogator modes. In half duplex mode, an interrogator can only send or receive information at on time, unlike a full duplex mode where both tasks can be performed at the same time.

handheld interrogator

A small, durable, and portable interrogator.

high frequency

3-30 MHz on the radio spectrum. HF RFID tags typically operate at 13.56 MH and can be read at 3 feet away.

host computer

Central hardware and network fixture that receives and process tag information.

impedance

The total opposition that a circuit offers to the flow of alternating current or any other varying current at a particular frequency. It is a combination of resistance, R, and reactance, X, measured in ohms.

inductive coupling

A method of transmitting data between tags and readers in which the antenna from the reader picks up changes in the tag's antenna.

inductance

The circuit property in opposition to the circuit's current flow.

input impedance

A test or measure resistance to the flow of a circuit's signal at the circuit input or load.

integrated antenna

An antenna that is configured with and built into another device, such as antennas used in wireless devices.

integrated circuit (IC)

A microelectronic semiconductor device comprising many interconnected transistors and other components. Most RFID tags have ICs.

International Organization for Standardization (ISO)

A global body that verifies standards for a wide array of field, including RFID.

interrogator

A device used to communicate with RFID tags. The interrogator has one or more antennas, which emit radio waves and receive signals back from the tag. The reader is also sometimes called an interrogator because it "interrogates" the tag.

interrogation zone

The area in which an interrogator successfully communicates with transponders.

interrogator/encoder

As called a read/write interrogator, interrogator/encoders have the ability to read and encode tags.

keying

The shifting of a data signal properties between a set of states.

line-of-sight transmission

Transmission when the transmitted and received antennas can physically see each other.

lobes

An identifiable segment of an antenna radiation pattern.

lossless power

Power that does not depreciate in quality

low frequency

From 30-300 kHz. Low-frequency tags typical operate at 125 kHz or 134 kHz.

magnetic fields

A region of space resulting from the current flow of the conductor.

medical implant

Some medical implants use short-range RF signals. For this reason, RFID system can interfere with such medical implants, which can be harmful to both humans and systems.

memory

The amount of data that can be stored in an RFID transponder's microchip.

microwave frequency

A segment of the ultrahigh frequency spectrum at 5.8 GHz.

middleware

A communications layer that allows applications to interact across hardware and network environments.

mobile mount interrogator

A interrogator that can be affixed to a forklift or other vehicle.

modulation

The process by which some characteristic of a higher frequency wave is varied in accordance with the amplitude of a lower frequency wave.

mono-static antenna

An antenna that both receives and sends signals. Interrogator with mono-static antennas can only operate in half duplex mode.

narrowband

A radio signal that occupies a small amount of space on the radio spectrum.

near field

The region (typically within one wavelength of an antenna) where the magnetic and electric fields are not related solely by characteristic impedance of free space.

noise

Any unwanted electrical signals or electromagnetic energy located in the operational RFID environment.

Object Name Server (ONS)

EPC designation for a local application system that organizes and locates data

omnidirectional antenna

An antenna that radiates signals in multiple directions.

operating frequency

Frequency produced by a component during operation.

panel antenna

A rectangular-shaped directional antenna designed to radiated signals perpendicularly and horizontally.

parabolic antenna

An antenna type (also called a dish antenna) utilizing a paraboloid- shaped reflector with a small, low-gain feed directional antenna.

passive tag

A transponder without an internal power source or battery.

phase shift keying (PSK)

A type of phase modulation using preset phase changes, contrary to relative phase changes, to represent digital symbols. PSK is usually performed by a modem for digital transmission on telephone lines.

phantom read

When a reader reports the presence of a tag that doesn't exist.

Physical Markup Language (PML)

An Auto-ID center-designed method of describing products in a way computers can understand. PML is based on the widely accepted eXtensible Markup Language used to share data over the Internet in a format all computers can use.

polarization

The direction of the electric field in the lightwave.

power gain

An increase in antenna or circuit signal power.

power level

The amount of energy transmitted from an interrogator to an active transponder.

radiation efficiency

The ratio of an antenna's total power radiated to its net power received.

radiation resistance

The aspect of an antenna's feedpoint resistance caused by the antenna's geometric propagation of electromagnetic waves.

radiation patterns

The radiation or antenna pattern describes the relative strength of the radiated field in various directions from the antenna, at a fixed or constant distance. The radiation pattern is a "reception pattern" as well, since it also describes the receiving properties of the antenna.

radiative coupling

Radiative coupling (backscatter) is deflection of a signal in the opposite direction of the signal.

read

The process by which data is retrieved from a transponder by a interrogator.

read-only tag

A RFID transponder that can only be read and is not reusable unless it is electronically reprogrammed.

read range

A term used to express the maximum distance from which an interrogator can maintain communication with a transponder.

read rate

A term used to express the maximum amount of information or number of transponders that can be read by an interrogator in a fixed period of time. The read rate is measured in bits per second.

read-write tag

A reusable RFID transponder that is able to store data.

Reader Talks First (RTF)

A transponder protocol for exchanges between the interrogator and a passive transponder. In RTF protocol the interrogator transmits the initial signals to activate the transponder.

reflection

The bouncing of a wave off of an object.

refraction

The bending of a wave as it passes from on medium to another, such as light bending thought water.

relative permittivity

Also called the dielectric constant, relative permittivity is a measure of a insulators ability to establish electric flux.

resonance

A circuit property where by the frequency of external forces matches a circuit's natural frequency.

return loss

A ratio of an antenna's transmitted power to reflected power caused by changes in impedance. Return loss is expressed in decibels and is an expression of VSWR.

RLC circuit

A electrical circuit composed of a resistor (R), inductor (L), and capacitor (C).

sector antenna

A large dish or panel antenna type used to transmit radio waves over large areas with precision.

semi-passive tag (battery-assisted tag)

A transponder that uses a battery to power its circuitry but not to transmit signals. Semi-passive tags are activated by a interrogator before they transmit data.

smart label

A label that contains an RFID transponder, which can store data such as a unique identifier (UID).

spectrum analyzer

A device used to measure RF signal, intensity, and captured power density.

spread spectrum

A system that spreads signal over a wider bandwidth than necessary to avoid interference.

standing waves

Waves in the transmission line that are stationary as the result of equal backward and forward moving opposition.

tag

Any radio transmitting and receiving device such as a smart label that is affixed to a object or product. Transponders are often activated by a predetermined signal from an interrogator.

tag talks first (TTF)

A transponder protocol for exchanges between the interrogator and an active transponder. In TTF protocol the transponder transmits a continuous signal, rather than responding to an activation signal from the interrogator.

transmitter

An electronic device connected to an antenna used for the generation of electromagnetic energy.

transceiver

A device that both transmits and receives radio waves.

transponder

Any radio transmitting and receiving device such as a smart label that is affixed to a object or product. Transponders are often activated by a predetermined signal from an interrogator.

voltage standing wave ratio (VSWR)

A ratio of an antenna's transmitted to received power in regards to a 50-ohm impedance line. A VSWR of one is 100 percent efficient.

ultrahigh frequency (UHF) range

The band of frequencies ranging from 300 MHz to 3 GHz. RFID devices on the UFH frequency range typically operate between 866–960 MHz.

Unique Identifier (UID)

A unique number used to identify the transponder.

Universal Product Code (UPC)

The barcode standard used by most retailers in North America.

write once, read many (WORM)

A tag that is can be written to once and is read-only.

write rate

The rate at which data is transmitted to a tag, written to the tag's memory, and verified by the transponder.

Yagi-Uda antenna

A directional antenna made up of one driven element and one, or more, parasitic elements

Appendix D
Equations

antenna efficientcy (4-2)

$$\eta = \frac{P_{rad}}{P_{in}} \times 100\ \%$$

Where $P_{in} = P_{rad} + P_{dis}$ (in Watts)

antenna gain (4-3)

$$D = \frac{P}{P_{ref}}$$

P = power density at a specific point with a given antenna (in Watts)

P_{ref} = power density of a reference antenna at the same point (in Watts)

antenna Q (4-12a)

$$Q = \frac{f}{f_h - f_l}$$

f = optimum frequency of operation

antenna Q for bandwidth (4-12b)

$$Q = \frac{1}{1.2 - 0.8} = 2.5$$

bandwidth (4-11)

$$BW = f_h - f_l$$

f_h = high half-power frequency

f_l = low half-power frequency

capture area (4-10)

$$A_c = \frac{A_r \cdot \lambda^2}{4 \cdot \pi}$$

A_c = effective capture area

P = wavelength of receive signal

A_r = power gain of receive antenna

captured power density (4-8)

$$C = \frac{P_{in} \cdot A_t \cdot A_r}{4 \cdot \pi \cdot R^2}$$

C = captured power density

P_{in} = transmit antenna input power

A_t = transmit antenna power gain

A_r = receive antenna power gain

R = distance between the transmit and receive antenna

Effective Isotropic Radiated Power (ERIP) (4-5a)

$$EIRP = P_{rad} \cdot D_t$$

P_{rad} = total radiated power (in Watts)

D_t = transmit antenna directive gain

EIRP dBm (4-5b)

$$EIRP_{(dBm)} = 10 \log \frac{P_{rad}}{0.001\ W} + 10 \log D_t$$

EIRP dBW (4-5c)

$$EIRP_{(dBW)} = 10 \log (P_{rad} \cdot D_t)$$

EIRP for antenna input power (4-6a)

$$EIRP = P_{in} \cdot A_t$$

P_{in} = total antenna imput power

A_t = transmit antenna power gai

EIRP for antenna input power dBm (4-6b)

$$EIRP_{(dBm)} = 10 \log \left(\frac{P_{in} \cdot A_t}{0.001\ W} \right)$$

EIRP for antenna input power dBW (4-6c)

$$EIRP_{(dBW)} = 10 \log \left(P_{in} \cdot A_t \right)$$

EIRP for transmit antenna gain (4-7a)

$$P = \left(\frac{P_{in} \cdot A_t}{4 \cdot \pi \cdot R^2} \right)$$

EIRP for directive gain (4-7b)

$$P = \frac{P_{rad} \square D_t}{4 \square \pi \square R^2}$$

P = power density

P_{rad} = power radiated from transmit antenna

A_t = transmit antenna power gain

D_t = transmit antenna directive power gain

R = distance from the transmit antenna

impedance (1-6)

$$Z = \sqrt{\frac{\mu}{\varepsilon}}$$

Where Z_0 = character impedance

μ = magenetic permeability of the mediun

ε = electric permeability of the medium

input Impedance (4-13)

$$Z_{in} = \frac{E_i}{I_i}$$

Z_{in} = antenna input impedance

E_i = antenna imput voltage

I_i = antenna input current

lossless power gain (4-4a)

$$A_p = D\eta$$

D = directivity

η = efficiency

lossless isotropic power gain (4-4b)

$$A_p = 10 \log (D\eta)$$

oscillation

$$f = \frac{1}{2\pi\sqrt{LC}}$$

period (1-5)

$$Z = \sqrt{\frac{\mu}{\varepsilon}}$$

Where Z_0 = character

radiation resistance (4-1)

$$R_r = \frac{P_{rad}}{i^2}$$

i = antenna current at the feedoint (in Amperes)

resonant frequency (1-2)

$$f_{res} = \frac{1}{2\pi\sqrt{LC}}$$

resonant frequency of RLC circuit (1-3)

$$f_{res} = \frac{1}{2\pi\sqrt{LC}}$$

return loss (4-14)

$$\text{Return loss} = 10 \log \frac{P_{reflected}}{P_{incident}}$$

wavelength (velocity, frequency) (1-4)

$$\text{wavelength} = \frac{\text{velocity}}{\text{frequency}} \quad \text{or} \quad \lambda = \frac{c}{f}$$

Appendix E
Figures

Appendix F
References

"Active Radio Frequency Identification (RFID) Powers Productivity and Protection in the Enterprise," white paper, Axcess Inc, 2005.

Aim Global. "RFID Standards." Aim Global Inc., 2001. http://www.aimglobal.org/standards/rfidstds/RFIDStandard.asp.

Bhuptani, Manish, Moradpour, Shahram. *RFID Field Guide*. Prentice Hall PTR, 2005.

Carr, Joe. "Directional or Omnidirectional Antenna?" Joe Carr's Radio Tech-Notes, 2004. http://www.dxing.com/tnotes.htm

Chang, K. *RF and Microwave Wireless Systems*. New Jersey: John Wiley and Sons, 2000.

"Choosing a Wireless Antenna," white paper, Black Box Corporation, Zaventem, Belgium, 2005.

Cole, Peter Harold, Ranasinghe, Damith. "Coupling Relations in RFID Systems" Auto-ID Centre, 2003.

Coleman, C. *An Introduction to Radio Frequency Engineering*. Cambridge: Cambridge University Press, 2004.

Collins, J. "Power Paper Gets Semi-Passive" *RFID Journal,* March, 2004.

Columbia Encyclopedia, 6th ed., s.v. "Transmission and Reception of Radio Waves." New York: Columbia University Press, 2005.

"Direct Sequence vs. Frequency Hopping," white paper, Wave Wireless Networking, 2000.

"Draft Protocol Specification for a 900 MHz Class 0 Radio Frequency Identification," Auto-ID Center, February 23, 2003.

"EPCglobal Class 1 Gen 2 RFID Specification," white paper, Alien Technology, 2005.

"EPC Generation 1 Tag Data Standards," Version 1.1, Rev 1.27, EPCglobal Inc., 2005.

"EPCglobal Class 1 Gen 2 RFID Specification," white paper, Alien Technology Corporation, Morgan Hill, CA 2005.

Finkenzeller, Klaus. *RFID Handbook*. 2nd ed. New Jersey: John Wiley and Sons, 2003.

Frenzel, Louis. "Tag It! From the feed lot to the parking lot, is this the year for RFID?" Electronic Design Online, May 12, 2005.

"Generation 2: A User Guide," white paper, ThingMagic, 2005.

"Guide to RFID Reader Selection," white paper, Intermec Technologies Inc., 2004.

Harrison, Mark., Hodges, Steve. "Demystifying RFID: Principles & Practicalities," white paper, Auto-ID Centre, 2003..

Integrated Publishing, *Introduction to Wave Propagation, Transmission lines, and Antennas*. Integrated Publishing, 2005. http://www.tpub.com/content/ts/14182/css/14182_13.htm.

"Introduction to Direct-Sequence Spread-Spectrum Communications," white paper, Maximum Integrated Products, 2003.

Jones, Edwin C, PhD, "Electromagnetic Waves," The Basics of Radio Wave Propagation. http://ecjones.org/propag.html.

Jordan, E.C., *Reference Data for Engineers: Radio, Electronics, Computer, and Comunications*. 7 ed. Indianapolis: Macmillan, 1986.

Kleist, Robert A. Sakai, David A., Jarvis, Brad S. *RFID Labeling: Labeling Concepts and Applications*, 2nd ed. Printronix, 2004

Knowlton, A.E., *Standard Handbook for Electrical Engineers*. 7 ed. McGraw Hill, 1941.

Kraus, John D., K.R.C., *Electromagnetics*. 2 ed. McGraw Hill, 1973.

Kuphaldt, T.R., Lessons in Electric Circuits. 2005. http://www.ibiblio.org/obp/electricCircuits/.

Leaver, S., "Evaluating RFID Middleware." Forrester, August, 13, 2004. http://www.for rester.com/Research/Document/Excerpt/0,7211,34390,00.html

Lillesand, Thomas M., Kiefer, Ralph W., Chipman, Jonathan, W., *Remote Sensing and Image Interpretation*. 4th ed. New Jersey: John Wiley, 2000.

Loy, Matthew, Karingattil, Raju, Williams, Louis. "2002 ISM Band and Short Range Device Regula tory Compliance," report, Texas Instruments, 2005

Magnusson, Philip C., Tripathi, Vijaj K., Weisshaas, Andreas, Alexander, Gerald C. T*ransmission Lines and Wave Propagation*. London: Taylor and Francis CPR Press, 2001.

"Next Up: Class 3 RFID." RFID News & Solutions, June 15, 2005. http://rfidnas.com/index.asp?layout=rfid_homeNew.

Powell, K., "Passive Radio Frequency Identification (RFID)–A Primer for New RF Regulations," white paper, Matrics Inc., 2003.

Purdie, Ian C., Antenna Basics. www.electronictutorials.com, 2000. http://www.electronics-tutori als.com/antennas/antenna-basics.htm.

Rau, Bill, "Brushing Up on VSWR," Wireless Review January 1, 1998. http://www.wirelessreview.com/ar/wireless_brushing_vswr/.

Reisert, J.H., "Understanding and Using Antenna Radiation Patterns," white paper, Astron Wire less Technologies, Inc., 2005.

Roberti, Mark. "EPCglobal Sets Up New Working Groups." RFID Journal March 3, 2005. http://www.rfidjournal.com/article/articleview/1427/1/1/.

Robbin, Allen H., W.C.M., *Circuit Analysis: Theory and Practice*. 2 ed. New York: Delmar, 2000.

Roberts, R. "Introduction to Spread Spectrum," The ABCs of Spread Spectrum–A Tutorial. Spread Spectrum Scene Online Inc., 2004. http://www.sss-mag.com/ss.html#tutorial.

Sarma, Sanjay, Engles, Daniel. "On the Future of RFID Tags and Protocols," Technical Report September 1, 2003. Auto-ID Centre.

"Securing Wireless Technology," white paper, Intermec Technologies Inc., 2005.

Shepard, Steven. *RFID: Radio Frequency Identification Technology*. McGraw Hill Professional, 2005.

Seidler, C., "RFID Opportunities for mobile telecommunication services," report, International Telecommunications Union, May 2005.

"Spread Spectrum Basic Summary," white paper, Airlinx Communications, 2005.

"Specification for RFID Air Interface." EPCglobal Inc., 2005.

"Specifications & Ratified Standards: Summary of Current Standards," white paper, EPCglobal Inc., 2005.

"Supply Chain RFID: How it Works and Why it Pays," white paper, Intermec Technologies Inc., 2005.

"Supply Chain Technology: Tracking to the Future," white paper, Bear, Stearns and Co. Inc., 2004.

Sweeney, Patrick J., *RFID for Dummies*. Indianapolis: John Wiley and Sons, 2005.

The Physics of Resonance. Intuitor.com, April 2, 1996. http://www.intuitor.com/resonance/.

Tomasi, Wayne. *Electronic Communications Systems*. 5 ed. Columbus: Prentice Hall, 2005.

"Understanding Radio Frequency Identification," white paper, R. Moroz LTD, 2004.

"Wave Propagation Along a Transmission Line," white paper, Agilent Technologies, Palo Alto, CA 2001.

Index

U

V

W

Y